Zscaler Cloud Security Essentials

Discover how to securely embrace cloud efficiency, intelligence, and agility with Zscaler

Ravi Devarasetty

BIRMINGHAM—MUMBAI

Zscaler Cloud Security Essentials

Group Product Manager: Wilson Dsouza
Publishing Product Manager: Rahul Nair
Senior Editor: Rahul Dsouza
Content Development Editor: Sayali Pingale
Technical Editor: Shruthi Shetty
Copy Editor: Safis Editing
Project Coordinator: Neil Dmello
Proofreader: Safis Editing
Indexer: Tejal Soni
Production Designer: Jyoti Chauhan

First published: May 2021

Production reference: 1070521

Published by Packt Publishing Ltd.
Livery Place
35 Livery Street
Birmingham
B3 2PB, UK.

ISBN 978-1-80056-798-6

www.packt.com

To the Supreme Lord, Sri Krishna, who is the driving force for every atom in the universe. To my wife, Lata, who inspired and encouraged me at every step of this adventure. To my children, who were very patient with me throughout this journey.

– Ravi Devarasetty

Contributors

About the author

Ravi Devarasetty is originally from India and came to the United States for his higher education. He started his IT career in embedded software development, moved into 24x7 network operations, later transitioned into secure web gateways, and now works in public cloud security. He likes constant learning, both through self-study and via mentoring relationships. He likes to tinker with technology and loves it when he is able to put the things he has learned toward creating a unique solution. He has experience working as a Zscaler solution deployment engineer as part of a **Managed Security Service Provider** (**MSSP**) and as a Zscaler consultant. He holds multiple Zscaler certifications, and is also certified in CISSP, CCSK, AlienVault, AWS, and Microsoft Azure.

About the reviewer

Anil Kumar Chennojwala is a seasoned security practitioner with a focus on the information security domain, comprising architecture, design, implementation, and service management. He has been helping customers spanning different verticals, from aviation to banking and finance, retail, and technology consulting. He hails from a small town called Karimnagar in the state of Telangana, India. Having received his master's degree in information systems security, he began his career as a network security engineer, and his enthusiasm for security got him into various IT and cybersecurity positions at companies such as United Airlines, Dell Technologies, Santander Bank, and Speedway LLC. Security is everyone's responsibility, and as such, his objective is to help foster that mentality across the technology industry and assist in developing solutions with a strong security architecture as a vital component of success.

I am eternally grateful to my ever-patient family for having supported me so much. A very special thanks to my managers, Eugene Silas, Johnny Kaissieh, and Rajneesh Bhambri, who have mentored me and helped me rise through the ranks to become who I am today.

Table of Contents

Section 1:
Zscaler for Modern Enterprise Internet Security

1

Security for the Modern Enterprise with Zscaler

2

Understanding the Modular Zscaler Architecture

3

Delving into ZIA Policy Features

4

Understanding Traffic Forwarding and User Authentication Options

5

Architecting and Implementing Your ZIA Solution

6

Troubleshooting and Optimizing Your ZIA Solution

Section 2: Zero-Trust Network Access (ZTNA) for the Modern Enterprise

7

Introducing ZTNA with Zscaler Private Access (ZPA)

8
Exploring the ZPA Admin Portal and Basic Configuration

9
Using ZPA to Provide Secure Application Access

10
Architecting and Troubleshooting Your ZPA Solution

Assessments

Other Books You May Enjoy

Index

Preface

Almost everyone in today's modern world knows about the internet and its role in everyday life, from email to live video calls. But at the same time, we have seen the rise of bad actors misusing and abusing the same internet for malicious purposes.

It is in this context that we need a secure way to browse the internet, especially from an enterprise perspective. **Zscaler Internet Access (ZIA)** is one such product that provides employees of the enterprise with a safe internet experience.

Many organizations have not yet embraced the concept of **Zero Trust Network Access (ZTNA)**. Most of them are still in a legacy security mindset. **Zscaler Private Access (ZPA)** provides ZTNA private application access to end users of the enterprise.

Who this book is for

This book is for a variety of readers. The first category consists of people like me who have been in the information technology field with no exposure to web security, but who want to transition into web security. The second category of readers is those who are in a decision-making capacity regarding a potential security product they are evaluating for their enterprise and are looking to compare Zscaler to other competing products. The third category consists of deployment and support engineers who need to architect, implement, and troubleshoot a Zscaler solution for an enterprise.

What this book covers

Chapter 1, *Security for the Modern Enterprise with Zscaler*, explains the evolution of the enterprise infrastructure and hence the need for unique, cloud-based, and scalable security solutions. It also introduces the two flagship products of Zscaler, namely, ZIA and ZPA.

Chapter 2, *Understanding the Modular Zscaler Architecture*, sets out the foundation for the reader by explaining the building blocks of the Zscaler cloud. It is very important to understand how the Zscaler cloud is architected in a modular fashion, and each component can scale without depending on the other components.

Chapter 3, *Delving into ZIA Policy Features*, gets right into the various web, mobile, and firewall features that are available with ZIA out of the box. A subset or all of the available features could be chosen by the Zscaler administrator of the enterprise.

Chapter 4, *Understanding Traffic Forwarding and User Authentication Options*, explains in detail the available options for forwarding the end user traffic to Zscaler. It also details the end user authentication options available to the enterprise and the process to choose the most appropriate option.

Chapter 5, *Architecting and Implementing Your ZIA Solution*, starts with the discovery of the current security posture within the enterprise, crafting a customized ZIA solution, and then implementing it across the enterprise.

Chapter 6, *Troubleshooting and Optimizing Your ZIA Solution*, provides practical troubleshooting tips for engineers supporting the ZIA solution and also provides ideas on how to get the most out of your deployed ZIA solution, such as reports and tweaking the dashboards.

Chapter 7, *Introducing ZTNA with Zscaler Private Access (ZPA)*, introduces the concept of and the need for ZTNA. It also explains how ZPA aligns with the fundamental principles of ZTNA, and lists the components of ZPA architecture and agentless ZPA solutions.

Chapter 8, *Exploring the ZPA Admin Portal and Basic Configuration*, takes readers through a tour of the ZPA admin portal, configuration of the ZPA log servers, end user authentication with Azure AD and Okta, and ends with the configuration options for the ZCC app.

Chapter 9, *Using ZPA to Provide Secure Application Access*, continues with the configuration elements of the ZPA solution, including App Connector deployment and application configurations, and explores best practices for enterprise deployments.

Chapter 10, *Architecting and Troubleshooting Your ZPA Solution*, walks you through the process of developing a customized ZPA solution and provides ZPA troubleshooting tips to the enterprise engineers supporting the solution.

To get the most out of this book

You will need access to either the ZIA Admin Portal or the ZPA Admin Portal or both in order to perform the configuration steps listed in this book. If you are administering end user authentication, it is recommended to have access to the IdP portal as well.

To set up the enterprise infrastructure, such as the NSS VMs or log servers, you need access to a VMware infrastructure within the enterprise.

Additional information regarding the hardware requirements is available on the Zscaler Portal.

Make sure to refer to the latest requirements from the Zscaler Help Portal, which can change frequently.

Download the color images

We also provide a PDF file that has color images of the screenshots/diagrams used in this book. You can download it here: `http://www.packtpub.com/sites/default/files/downloads/9781800567986_ColorImages.pdf`.

Conventions used

There are a number of text conventions used throughout this book.

`Code in text`: Indicates code words in text, database table names, folder names, filenames, file extensions, pathnames, dummy URLs, user input, and Twitter handles. Here is an example: "Zscaler's advanced sandbox feature can scan many additional types of file, such as `.doc(x)`, `.xls(x)`, `.ppt(x)`, and `.pdf`."

Bold: Indicates a new term, an important word, or words that you see on screen. For example, words in menus or dialog boxes appear in the text like this. Here is an example: "Click **Flash** from Etcher to write the image."

> **Tips or important notes**
> Appear like this.

Get in touch

Feedback from our readers is always welcome.

General feedback: If you have questions about any aspect of this book, mention the book title in the subject of your message and email us at `customercare@packtpub.com`.

Errata: Although we have taken every care to ensure the accuracy of our content, mistakes do happen. If you have found a mistake in this book, we would be grateful if you would report this to us. Please visit `www.packtpub.com/support/errata`, selecting your book, clicking on the Errata Submission Form link, and entering the details.

Piracy: If you come across any illegal copies of our works in any form on the internet, we would be grateful if you would provide us with the location address or website name. Please contact us at copyright@packt.com with a link to the material.

If you are interested in becoming an author: If there is a topic that you have expertise in, and you are interested in either writing or contributing to a book, please visit authors.packtpub.com.

Reviews

Please leave a review. Once you have read and used this book, why not leave a review on the site that you purchased it from? Potential readers can then see and use your unbiased opinion to make purchase decisions, we at Packt can understand what you think about our products, and our authors can see your feedback on their book. Thank you!

For more information about Packt, please visit packt.com.

Section 1: Zscaler for Modern Enterprise Internet Security

In this part, you will learn about the need for security and how it needs to change as the modern enterprise and workforce evolves.

This section comprises the following chapters:

- *Chapter 1, Security for the Modern Enterprise with Zscaler*
- *Chapter 2, Understanding the Modular Zscaler Architecture*
- *Chapter 3, Delving into ZIA Policy Features*
- *Chapter 4, Understanding Traffic Forwarding and User Authentication Options*
- *Chapter 5, Architecting and Implementing Your ZIA Solution*
- *Chapter 6, Troubleshooting and Optimizing Your ZIA Solution*

1
Security for the Modern Enterprise with Zscaler

In the past few years, there has been a momentous shift in the way modern enterprises have evolved. They have moved from a traditional hub-and-spoke, data center type of network to a cloud-based or anywhere-access type of network. The core locations have become more decentralized because the employees are now based in various geographies and the applications are migrating to the cloud.

When we look at the infrastructure itself, enterprises invest in a variety of products such as routers, switches, and firewalls to implement various functions such as authentication and security. These products very quickly reach end-of-life from a capacity and a vendor-support perspective. This, in turn, causes the enterprises to upgrade in a 3- to 5-year cycle where they must do a lift and shift of the entire hardware in their data center. This moves the enterprise expenditure from an OPEX to a CAPEX model, which is not desirable from a business and planning perspective.

In this chapter, we will see how Zscaler steps in as a cloud-based security solution. The ZIA product provides secure internet access and the ZPA product brings the geographically spread-out end users and enterprise applications together. They both provide the following benefits:

- There are no upgrade cycles for the enterprise as Zscaler takes care of that.
- There is a shift from CAPEX to OPEX, which enterprises like because of predictability.
- An amazing user experience as users can access applications using the best path.

In this chapter, we are going to cover the following main topics:

- Fundamental definitions in security
- Shift of the modern enterprise and its workforce
- The need for scalable, cloud-based security
- **Zscaler Internet Access (ZIA)** for a safe and secure internet experience
- **Zscaler Private Access (ZPA)** for a zero-trust private application access

Let's get started!

Fundamental definitions in security

In this section, we will define some commonly used internet and security terms that are applicable to this book. A detailed explanation of all internet and security concepts is outside the scope of this book. If you are already comfortable with these terms, you can skip ahead to the next section.

Active Directory

Active Directory is a directory service that was originally developed by Microsoft for the Windows environment and was released in 2000. It stores data such as users, groups, and devices. It has many components that assist the user to interact with the domain. Our focus in this book is to authenticate users against their credentials in Active Directory.

Authentication

Authentication is the process by which an end user, a computer, or a software application can prove its identity. This is typically done using a username and a password. The term **multi-factor authentication** (**MFA**) is gaining popularity today. MFA means that there is an additional item that is needed in addition to a username and a password. This could be a token number or a biometric such as a fingerprint or a retina scan.

Bad actors

A bad actor is, in general, a malicious party that is usually interested in the following:

- Attacking legitimate users and businesses due to various motivations
- Stealing sensitive and valuable information from individuals and businesses
- Compromising infrastructure such as servers and using them for their needs

Next, we'll look at bandwidth.

Bandwidth

Bandwidth refers to the rate of data transfer over a network. It is typically measured in bits per second. The higher your bandwidth, the faster you can transfer your data across. The data being transferred could be an image, text, a video, or a combination of all three.

Certificate

A **certificate** is usually a small text file that can be used to establish the identity, authenticity, and reliability of a web server on the internet. Certificates are usually used to assure the confidence of end users trying to use the services of a website and to provide protection against malicious websites. Certificates are issued by certification authorities and they are usually tracked with creation and expiry dates.

DLP

Data Loss Prevention (**DLP**) is the prevention of loss of any kind of valuable or sensitive data. Valuable data may mean company proprietary formulas and business strategies. Sensitive information may be customer information such as social security numbers, credit card numbers, date of birth, and so on.

DNS

The **Domain Name System** (**DNS**) is a system that converts domain names (such as www. google.com) into IP addresses so that web browsers can translate customer requests into lower-level IP packets and carry on data transfer tasks, such as loading websites. The DNS is very crucial for internet security as bad actors can hijack these servers and have the end user traffic sent to their malicious web servers, instead of the legitimate ones.

Firewall

A **firewall** is a security device or application that monitors traffic through the network and applies security rules configured by the administrator to that network traffic. Firewalls are usually used as perimeter security devices by many organizations.

FTP

The **File Transfer Protocol** (**FTP**) is a network protocol (based on IETF standards) that is used primarily to transfer files between a client and a server across a network.

Identity Provider

An **Identity Provider** (**IdP**) is a system that creates and maintains identity information for end users or applications. When a company wants to authenticate an end user, they usually make a call to the IdP. An IdP is essentially an **Authentication as a Service** (**AuthaaS**).

Intrusion Prevention System

An **Intrusion Prevention System** (**IPS**) is a system that sits in the line of the network traffic and looks at possible malicious activity and blocks it. There are many types of IPS systems, with the most recent ones looking to leverage artificial intelligence and machine learning.

Kerberos

Kerberos is an authentication protocol used on computer networks. It issues tickets for end user access and allows end points to communicate over non-secure network systems, and then prove their identity to one another in a secure way.

Logging

In the security world, **logging** means to record the transactions going across the network to a file on a storage medium. When there is a need to investigate a security incident, these logs are then analyzed by specialized systems to derive insights and conclusions.

Malware

Usually, software applications are used for legitimate purposes, such as for operating and growing a business. But bad actors write malicious software with the intent to steal valuable information or attack infrastructure such as computers. This malicious software is called **malware**. It could be as damaging as bringing down an entire organization to its knees or as annoying as pesky advertisement popups.

PAC file

Usually, individuals sitting at their home computer access the internet directly. But many organizations use a proxy server that sits between the end users and the internet. They do this to monitor their employees' activity against any company policy violations. A **proxy auto-config** (**PAC**) file defines what proxy servers and methods are chosen by end user web browsers. A simple example would be choosing ProxyServer1 when going to www. yahoo.com and choosing ProxyServer2 when going to www.google.com.

SAML

Security Assertion Markup Language (**SAML**) is an open standard that is used to exchange authentication and authorization information between an IdP and a service provider. For example, some websites allow you to log in using your Google account. End users navigate to the website of interest. They click on **Sign in with Google** and are then redirected to Google. The user then enters their Google credentials, and they are authenticated and are then redirected to the original website. In this case, the original website is the service provider and Google is acting as the IdP.

Sandbox

A **sandbox** in security is an isolated environment where software components may be executed to observe their behavior and note down any malicious intent. Unknown software components are typically "detonated" in a sandbox environment before they are passed on to the end user.

Secure Web Gateway

A **Secure Web Gateway** (**SWG**) is a component or solution that continuously monitors web traffic between end users and web servers, and filters any traffic that is malicious or does not comply with the enterprise policies.

Secure Sockets Layer/Transport Layer Security

Secure Sockets Layer (**SSL**)/**Transport Layer Security** (**TLS**) are cryptographic protocols that provide secure communication over a typically untrusted connection or network. They are commonly used when exchanging sensitive information, such as typing in your social security number or a credit card number on a website. Your browser typically shows a "*lock*" icon just in front of the URL in the address bar.

Surrogate IP

When an end user types in their credentials and are authenticated, a relationship is established between that user and the IP address they are currently using to access the network. This assumes that the IP address is used by only one user within the entire organization at any given time. So, this IP address is treated just like the user in terms of granting access to applications and so on.

Tunnel

When using an untrusted network such as the internet, private communications can typically be placed inside of (encapsulated) other packets. This allows for data to be moved across the untrusted network securely. This process is called tunneling. The channel that is established for this purpose is called a tunnel. There are many types of tunnels, such as GRE, IPSec, and so on.

VPN

A **Virtual Private Network** (**VPN**) allows an enterprise to extend their private network across a public network. For the end users, it appears as if the other side of the network is right across the room.

XFF

When an end user connects to a website through a proxy, the proxy will put its IP address when communicating with the web server. The **X-Forwarded-For** (**XFF**) header field can be used to identify the IP address of the originating end user. It can be extracted by the web server to make decisions based on the originating IP address of the end user.

With that, we have briefly touched upon the basic technologies that you will encounter in this book. Though this was a brief introduction, in this book and in your own work, you will get to know many of these concepts in more detail. In the next section, we will explore the changes that have led to the modern enterprise and workforce that we know today.

Understanding the evolution of the modern enterprise and its workforce

In this section, we will learn how the modern enterprise has slowly moved away from a central data center or headquarters model to a more distributed, internet-based model. We will also learn how the working habits of the enterprise workforce have changed with the advent of working remotely over the internet.

Evolution of the workforce

With the advent of the internet, for many technology workers, what could be done in the office can now be done remotely over the internet using technologies such as VPNs. This shift was accelerated due to several reasons:

- Employees want a flexible work style. They no longer are tied to a traditional 8 A.M. to 5 P.M. work schedule.

- Various teams in the companies now make up employees from different geographies, so 8 A.M. is no longer the same for everyone on the team.

- Companies benefited by moving from a dedicated office space (such as a cubicle for an employee and an office room for a manager) to a flexible workspace. This way, there are some flexible workspaces that could be reserved by the employees on the days where they want to come to the office.

- Different roles for the employees mean that someone could be working on a production install after-hours, which is better from the comfort of their home than a lonely work location with no one around.

- With the COVID-19 pandemic raging across the world, employees do not want to put their families at risk, and the pandemic has accelerated the move to work remotely over the internet.

All these points mean that now, companies must adapt to their workforce. They must make applications readily available to their employees wherever they are located.

Enterprise infrastructure evolution

In a data center architecture, the enterprise chooses certain locations to serve as their repositories for applications and data. A company may choose a certain city on each of the continents they operate in and provision and maintain a massive data center. At this point, the company needs to provision expensive private connections between all its offices and these data centers.

Very quickly, this becomes an expensive proposition for the company. Not only does it have to focus on its core business, but now it must run and maintain its massive infrastructure. This infrastructure consists of several product categories, such as routers, switches, firewalls, and application servers. For redundancy and high-availability purposes, the company must invest double the amount of equipment, even if the chances of a failure on the hardware components is low. This is because it cannot take the risk of business application downtime.

To add to this complexity, we all know that hardware for these products quickly becomes out of date. We are all familiar with our own personal upgrade cycles where we upgrade our electronic gadgets such as our smartphones, laptops, and tablets. Corporations are in a similar upgrade situation every 3 to 5 years based on the manufacturer, the product, and the technological changes in the marketplace.

When these upgrades come around, there is a wholesale lift-and-shift of the entire hardware, which needs a lot of manpower. This upgrade is also treated as a **capital expense** (**CAPEX**) and not as an **operating expense** (**OPEX**). Enterprises prefer an OPEX model because it allows them to predict the costs and account for them in their business operating model.

Enterprises also have a range of products doing different things. Most of the time, they do not have a choice, even if one product overlaps with another product in terms of its features. There is no single magic bullet or integrated product that can meet all the customers' needs.

Now that we've learned about the evolution of the preferences of the enterprise workforce and the changing requirements for the enterprise infrastructure, let's look at how a cloud-based security solution can address both those needs.

Exploring the need for scalable, cloud-based security

In this section, we will see how these shifts in trends lead us toward a scalable, available, cloud-based security while using the internet as the underlying transport mechanism.

Workforce evolution requirements

As the workforce evolves and demands access to applications from anywhere, we must look at the common medium of transmission. We can all safely agree that the internet seems to be that common medium. End users can now access the internet using several methods such as a computer (Ethernet), a tablet (Wi-Fi), or a smartphone (cellular network). The internet is now considered a utility like electricity, water, and gas. So, why not use the internet to bring these end users to their applications?

The workforce is also demanding access not only from anywhere but at any time. Again, the internet solves this problem. The internet is always on. Many **Internet Service Providers (ISPs)** now provide **service level agreements (SLAs)** like other utilities.

Enterprise preferences

Now, let us look at what we need in order to develop a model that enterprises prefer. The first issue was trying to build a vast network and infrastructure to host their applications and then to connect their workforce to those applications. If enterprises were to leverage the universal medium – the internet – they could use it as the transport mechanism to connect their workforce to their applications. This is very much true for internet-based applications, but it could also work for in-house legacy applications that run on physical servers.

Enterprises could migrate their applications to virtual servers on various public cloud platforms such as **Amazon Web Services (AWS)**, **Microsoft Azure**, and **Google Cloud Platform (GCP)**, or they could somehow leverage the internet to connect their users to the legacy applications in their data centers.

The second problem is the constant, expensive upgrade cycle. What if the provider is cloud-based and all upgrades are managed by the provider without any burden on the enterprise? All the enterprise needs to do is hand off their traffic to the provider using the internet; the provider does the rest. The enterprise is guaranteed a SLA from the provider and is also provided with high availability. This model also shifts the spending model from CAPEX to OPEX, which is preferable by the enterprise.

The third problem is in terms of the various products needed for a set of features. What if the enterprises can rely on a provider that has all the essential features that enterprises need and can be chosen on a subscription basis? Enterprises get the essential features for a base pricing model (billed monthly) and they can choose optional features for extra money. For example, they may choose extra features 1, 3, and 4 and pay $X more or choose extra features 1, 2, 3, and 4 and pay $Y more. Even better, what if these license costs are based on the number of active users? If an enterprise has 500 users, it pays 500X monthly instead of an arbitrary monthly amount. This would be a very fair pricing model, no different than a utility billing such as electricity, water, and gas.

Scalable, highly available, cloud-based solutions

Any security solution that is designed for enterprises needs to tick these boxes. A scalable solution means that the solution should continue to work at the same expectation levels when the user count goes from 100 to 10,000. This provides assurance to the enterprises that they do not have to worry about poor performance as their user base scales up or down.

The solution also needs to be highly available. This means that when a certain component of the provider goes down, end user traffic should automatically be handled or re-routed by another component that is ready and standing by. The availability of the provider is usually measured using SLAs. Some SLAs that are often mentioned by providers are 99.99% available or 99.95% available.

Finally, enterprises prefer a cloud-based solution where they do not have to do or know anything about how the providers operate. All the enterprises do is forward their traffic to the cloud provider and that is the end of it. The cloud provider provides the enterprise with an administration portal where the enterprise administrators can log in and provision their desired configuration.

Internet security for everyone

In today's world, we are seeing that a lot of small businesses, schools, and hospitals are being targeted by bad actors, especially using ransomware that has been on the internet for quite some time. The consequences of a compromise can be fatal to these organizations. In the past, it was difficult to select and provision a security solution.

It does not have to be like that today. The solution that will be presented in the next sections is quite easy and quick to implement, especially when using the default security policy that is based on industry standards. This is even more true for a startup or a consulting organization that has many employees remotely working across broad geographies. As the saying goes, "*prevention is better than cure*" – this is very much true for internet security today.

Using Zscaler Internet Access for a safe and secure internet experience

The internet today has become the wild, wild, west. There is a mushrooming of many types of websites, especially after the dot com boom. It has become difficult to keep track of legitimate websites versus malicious ones. When the **Internet Service Providers** (**ISPs**) themselves cannot keep track of these harmful websites, we cannot expect the end user to keep up with it. This is why we need a security solution to give the end users a safe internet experience.

Why safe internet?

Employees of the enterprise have a business need to access the internet on an almost daily basis. This could be for researching solutions, learning new skills, or to log into internet-based applications for company work.

Employees may be directed to go to a website through various means. For example, they may receive an email with a link where they can access the latest content on an interesting topic. A friend or a co-worker could send a web link through an instant chat message.

When employees are using corporate-issued devices to access these websites, it is the duty of the enterprise to provide employees with safe and secure internet access. If the employees inadvertently access malicious websites and those websites install some sort of malware on the corporate-issued device, then that malware could spread to other enterprise systems, including critical infrastructure, which will have a massive impact on the enterprise.

This is no different than someone catching a viral infection and then going around spreading it inadvertently – hence the need for safe internet. For example, an employee receives a seemingly legitimate email telling them they can find more information on a topic at www.help.com. A spammer or a bad actor can easily change the letter "l" in the website URL to the number "1" so that the malicious URL is www.he1p.com. Based on the font used by the employee's email program, the difference may not even be that visible.

The employee then proceeds to click on the malicious link, thereby triggering the malware and compromising the machine. Internet security is needed because not all malicious emails may be caught by the company's email security software. This is where **Zscaler Internet Access (ZIA)** comes in.

How ZIA works

ZIA is a cloud-based web proxy whose primary purpose is to provide safe and secure access to the internet. Simply put, ZIA sits between the end user and the target internet website resource. The enterprise will purchase the necessary subscription and internet security feature set as part of their contract. A company Zscaler administrator will provision and activate these security settings in the ZIA portal. Those changes take effect immediately.

Once this has been set up, suppose an employee receives an email with a malicious link in it, as described in the previous section. When the employee clicks on that link, the browser on the machine tries to navigate to that malicious website. But that initial website request is now intercepted by Zscaler. Zscaler then checks this URL against its dynamic list of malicious websites and identifies it as a malicious website. Zscaler will then display a warning message that says this is a malicious website and hence the request was blocked.

A very impressive feature of ZIA is that it can detect botnet callbacks. Although we will talk about it in more detail in later chapters, we will provide an example here. Let's say that an employee takes their corporate device home and then accesses the internet in an insecure way, so the bot is now installed on their device. When the employee uses the same device in the Zscaler-protected corporate environment, Zscaler will identify and block that botnet callback to the central bot server and can also alert an administrator. The administrator can then immediately identify the device and the user, and then either quarantine that device or get it cleaned immediately using anti-malware software, thereby eliminating the root problem and preventing it from spreading. This can be visualized with the following diagram:

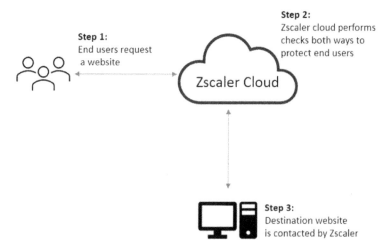

Figure 1.1 – Fundamental operation of Zscaler Internet Access (ZIA)

ZIA is also famous for its cloud sandbox feature. When malware is initially released on the internet, its signature (the bit pattern in binary) is not known to many anti-malware engines. ZIA can (adding a little bit of delay) identify this unknown signature and detonate it safely in its cloud sandbox environment and observe its effects. If there is no fallout, ZIA will forward that packet normally. If, however, it is observed that the malware is harmful, ZIA will immediately update its threat signature database and propagate that information to all its clouds, thus protecting all the remaining customers within a matter of minutes.

There are many ways ZIA can be provisioned. If a user is at a corporate location, GRE or IPSec tunnels can be established from the location to the two (there could also be more or less than two, depending on the customer's choice) nearest Zscaler cloud locations. If the user works remotely or travels a lot, an application called the **Zscaler Client Connector** (**ZCC**) can be installed on the user's device. Before the user can access the internet, the user will have to log into the ZCC using their credentials manually or by using their Active Directory Domain credentials. This makes sure the user is always protected.

Zscaler estimates that over 80% of the traffic on the internet is now using SSL. Hence, SSL inspection is an integrated, most basic feature that is supported by ZIA.

Using Zscaler Private Access for secure application access

Employees of the enterprise primarily work on the company applications that generate revenue, support customers, and grow the company business. These company applications have traditionally been custom-designed for the enterprise and hosted in data centers. With the expansion of the internet and public cloud providers, many enterprises are migrating their applications to the cloud. Employees need to securely connect to these applications in an effortless manner. Here, we will introduce the concept of private access.

What is Private Access?

In the previous sections, we looked at the security needed when the users are accessing the public internet. Many enterprises host their core business applications in a private data center or in the public cloud. Most of the time, the employees work on these business applications as part of their daily work duties.

In the past, we saw that most company employees go to their corporate location, access the business applications using their internal LAN and desktops, and then go home at the end of the day. But as we explained in the workforce shift, the following happened:

- Business applications started to move to the cloud (web-based model).

- Employees wanted flexibility in terms of where and when they worked.

- The internet became the most popular and affordable transport platform.

Now, enterprises can't force their employees to go to a corporate location anymore. So, how do enterprises connect end users to the business applications without exposing either to the public internet? This is where **Zscaler Private Access** (**ZPA**) comes in.

How ZPA works

The primary use case for ZPA is to connect the end users and the business applications wherever they are, without even traversing the public internet. What ZPA does is use the internet as a transport medium and heavily leverages the Zscaler cloud.

While installing the service, an enterprise ZPA administrator does the following:

1. End users are identified based on their department or location.

2. Business applications that need to transition to ZPA are identified.

3. The end user access policy to business applications is created using business needs.

4. A small virtual machine called the App Connector is deployed near the business applications.

We can see this illustrated in the following diagram:

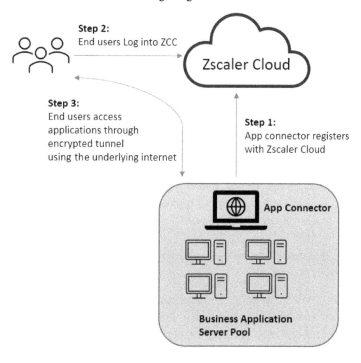

Figure 1.2 – Fundamental operation of ZPA ADD

When the App Connectors boot up, they discover the business applications and register with the Zscaler cloud, stating that they are ready to serve the end users to provide access to end users. This communication happens over secure tunnels using the internet as the underlying transport mechanism.

When users log into their corporate devices, they authenticate with the ZCC application, as described earlier. Based on *Step 3*, their application access is provisioned and ready upon authentication, and this happens in a transparent manner for the end user. The end users do not have to do anything special.

When the end users initiate the business application natively or using a web-based interface, their request is handled by the nearest Zscaler cloud. The Zscaler cloud already knows where that business application resides from App Connector registration. The Zscaler cloud then brokers the connection between the end user and the App Connector in the most optimum and secure manner.

As you can see, from an end user perspective, all they must do is log into ZCC, and everything just works! The end users or the applications are never exposed to the internet. You cannot attack what you cannot see!

ZCC

We saw ZCC mentioned in both the previous sections, so let's clarify things here. ZCC can be used for just ZIA, just ZPA, or both ZIA and ZPA. Ideally, the enterprise would use the same authentication mechanism so that their end users do not have to log into ZCC twice – once for ZIA and once for ZPA – which would be very confusing. ZCC is a central tool in most Zscaler implementations and as we will see in later chapters, ZCC offers a lot of different configurations that provide flexibility to each enterprise situation.

Summary

In this chapter, we saw how the internet has changed the working habits of the modern workforce influenced how enterprises operate. We learned about the need for a cloud-based, scalable, security solution. We then examined how Zscaler's ZIA and ZPA products play a key role in providing end users with a safe and secure internet browsing experience, while also providing them with secure access to the company's applications without being exposed to the internet.

In the next chapter, we will learn about the essential components of the Zscaler cloud, the roles and functionality of each component, and how they interact with each other. We will introduce the concepts of the management plane, the data plane, and the statistics plane.

Questions

As we conclude, here is a list of questions for you to test your knowledge regarding this chapter's material. You will find the answers in the *Assessments* section of the *Appendix*:

1. What are the reasons that a modern workforce prefers a flexible and remote work style?

    a. The internet can now allow anyone to access anything from anywhere.

    b. Flexible work style and globalization means no fixed hours or time zone.

    c. The enterprises require that their workforce is remote only.

    d. a and b

2. Why do enterprises prefer an OPEX rather than a CAPEX model?

    a. OPEX is better for accounting and tax purposes for enterprises.

    b. OPEX provides cost predictability, whereas CAPEX does not.

    c. All the above.

    d. None of the above

3. Why are enterprises moving away from the central data center model?

    a. Enterprises must keep up with the constant device upgrade cycle.

    b. Enterprises must connect all their data centers together with high-speed communications.

    c. Enterprises can achieve economies of scale by leveraging the public cloud.

    d. All the above.

4. The internet remains a safe place and hence there is no need for a web security solution for enterprises.

    a. True

    b. False

5. ZIA provides secure, private access to enterprise applications.

    a. True

    b. False

6. In a ZPA solution, neither the end users nor the enterprise applications are exposed to the internet.

    a. True

    b. False

2
Understanding the Modular Zscaler Architecture

In this chapter, we will introduce the modular and highly available architecture of Zscaler. Both **Zscaler Internet Access** (**ZIA**) and **Zscaler Private Access** (**ZPA**) use the same core Zscaler infrastructure, the only difference being the interaction of the various components for their specific purpose. The focus here will be the ZIA architecture, but the ZPA architecture is covered in detail in *Chapter 7, Introducing ZTNA with Zscaler Private Access (ZPA)*.

Zscaler has several clouds and has been adding more, as dictated by customer demand. Each Zscaler cloud consists of the **Central Authority** (**CA**); **Zscaler Public Service Edges** (**PSEs**), previously called **Zscaler Enforcement Nodes** (or **ZENs**); and **Nanolog clusters**. We will also mention Zscaler's **Single-Scan Multi-Action** (**SSMA**) technology that makes it so efficient and offers superior performance.

In this chapter, we are going to cover the following main topics:

- Introducing the Zscaler cloud architecture
- Understanding the CA—where the core resides

- Using Zscaler PSEs—where the policies are applied and enforced
- Monitoring user and application activity using Nanolog clusters
- Protecting enterprise users and infrastructure with Sandbox

Introducing the Zscaler cloud architecture

Let's get started with an overview of how the Zscaler cloud is architected to be modular and highly available, which is nowadays a bare-minimum necessity for enterprises.

When an enterprise is provisioned on a Zscaler cloud, they get an instance on that Zscaler cloud. The enterprise administrator then proceeds to customize and configure their security policies and controls in the assigned cloud instance. This configuration includes users, groups, departments, and a collection of management policies and settings. This information is used by the Zscaler component to act on the user traffic and enforce the policies.

The aforementioned configuration resides in the Zscaler cloud CA, which is the core that has the intelligence to manage the entire cloud. It is also the same engine that supplies the necessary information to other Zscaler cloud components to perform their functions. The CA is not just one server somewhere in the cloud; it is designed to be highly available and, hence, it is stored in multiple data centers for redundancy. Since this information is used to manage and control the entire cloud, this plane is called the **management plane** or the **control plane**.

The next component is the Zscaler PSEs, which are the *worker* nodes. When user data packets are flowing through the Zscaler cloud, the PSE nodes look at each packet and identify the customer, the policy to be enforced, the configuration, and so on, and then act accordingly on that packet or packet stream as applicable. Since these nodes work on the data packets, they are said to be in the data plane.

The components of the Zscaler cloud can be seen in the following diagram:

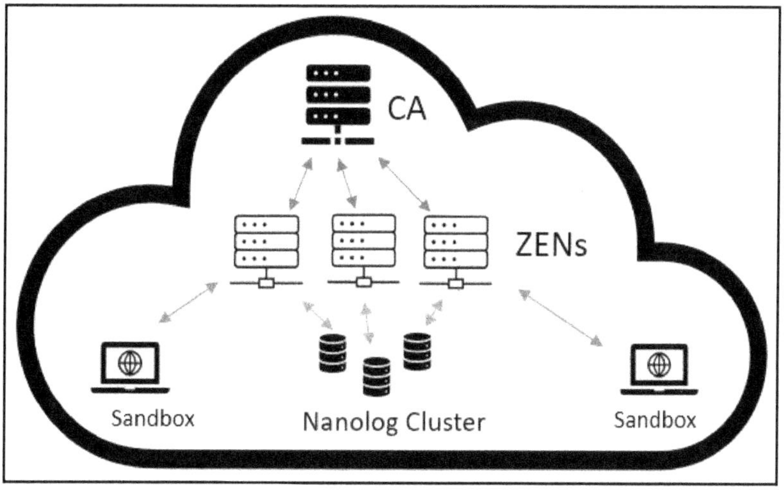

Figure 2.1 – Components of the Zscaler cloud

The Zscaler cloud also maintains a count of several types of packets, with details of total packets processed, number of malicious packets, number of packets dropped, and so on. Since this area deals with statistics within the cloud, this is called the **statistics plane**. Data logs are stored for up to 6 months by this statistics plane in a geographical location of the customer's choice (Americas, Europe, Asia-Pacific, and so on) and can also be optionally streamed to a customer location, to be consumed by a **Security Information and Event Management (SIEM)**.

Now that we have an understanding of the Zscaler components, let's see which technology gives Zscaler its high efficiency and performance.

SSMA

Although not a component in itself, it is worth mentioning Zscaler's SSMA technology. When it comes to scanning and processing data packets, many vendors use a concept called a **service chain**. What this means is that first, the **Uniform Resource Locator (URL)** filtering acts on the packet. It is then handed off to the anti-virus process, then to the anti-malware process, and so on. Each step adds an additional latency, slowing down the overall process.

Instead of service chaining, Zscaler uses its SSMA technology. In this case, a packet is placed into a large, shared memory. There are individual **central processing units** (**CPUs**) dedicated to performing a specific, specialized function—for example, one CPU will do the proxy function; one will perform **network address translation** (**NAT**); one will perform anti-virus; and so on. When all these CPUs are done with their processing at the same time, the overall result is calculated for that data packet, and a decision is reached on whether to allow or block the data packet.

An overview of the Zscaler SSMA technology can be seen in the following diagram:

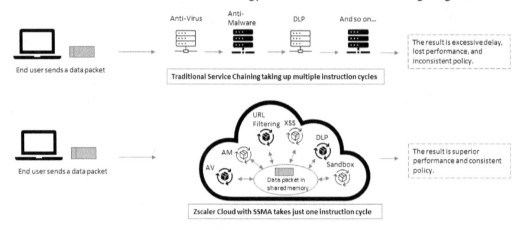

Figure 2.2 – Zscaler SSMA technology delivers performance and consistency

What happens when a customer purchases extra features not included in the standard subscription? Does it add to the latency because more processing is needed for those extra features? The answer is no. All the individual engines still run on the data packet, regardless of the customer's feature subscription.

This information may appeal to hardware lovers. Each CPU is performing the same function again and again, so it almost works like an **application-specific integrated circuit** (**ASIC**). The CPU is retiring an instruction every clock cycle and it never uses long-term memory or a disk. This means that the performance is many times faster because there are no interrupts to deal with.

High availability and redundancy

As with the major public cloud providers such as **Amazon Web Services** (**AWS**), Microsoft Azure, and **Google Cloud Platform** (**GCP**), Zscaler is also designed with high availability and redundancy at every level.

Zscaler has a built-in N+1 or N+2 redundancy. All inbound user traffic goes to a **virtual IP (VIP)** that is load-balanced N+1 into a redundant architecture. So, even if one or two instances went down at a data center level, it would not make much of a difference to performance or availability. If all the instances were to fail for some reason, then the failover would occur to a different data center.

For logging and reporting purposes, Zscaler uses N+2 redundancy. This means that the logs from the statistics plane are written and stored in three copies. So, even if two data centers were to go down, reporting and logging would still be operational.

Inside the data center, Zscaler is provisioned with component-level redundancy. The load-balancing architecture is software-based. Many software instances are created within a single hardware box and traffic is load-balanced across them. If a software instance fails, then it is simply restarted without restarting the entire hardware component. This is very similar to running several **virtual machines (VMs)** on your PC. A failure of one VM does not impact other VMs running on the computer.

Zscaler also has built-in redundancy for storage and power. A **Redundant Array of Inexpensive Disks (RAID)** is used for storage, and multiple power supplies are in place to provide power redundancy. This is the reason Zscaler offers a five-nines **service-level agreement (SLA)**, so important for modern enterprises.

In this section, we learned about the high-level components of the Zscaler cloud and how Zscaler's SSMA technology delivers superior performance. We also learned about how the Zscaler cloud is architected for high availability and redundancy.

Let's now explore in detail the various components of the Zscaler cloud that were introduced in the first section. Let's start with the first component: the CA.

Understanding the CA – where the core resides

As mentioned earlier, the CA is where an entire company's data will be stored, relating to users, groups, and departments; configuration such as access policies detailing who will be allowed access to what and when; where and how logging will be performed.

When a new customer is provisioned on a Zscaler cloud after a contract signature, the main contact on the contract is given a *Super Admin* credential (username and password) into the cloud instance that is created for that customer. The main customer contact can then use those credentials and log in to the Zscaler administration portal for their cloud instance. The customer will then have to decide how to translate their company security policy and configure the Zscaler administration portal that resides in the CA.

After the customer completes this configuration portion, the CA never uses customer information such as username, location, and company name as it is. Instead, the CA tokenizes this information and assigns a randomized unique identifier to each data entity of the customer.

For example, John Doe of Company ABC will have an identifier of 123456. This is done so that the other components of the Zscaler cloud—namely, the PSEs and Nanolog clusters— only know this identifier 123456, and not John Doe of Company ABC. That way, if a Nanolog cluster were to be breached by a bad actor, the other components do not have the reverse-translation dictionary that maps these random identifiers to a name, group, or department of the company.

From a location perspective, the CA does not reside in every Zscaler data center. The CAs reside in at least three to four data centers, which is more than enough from a disaster-recovery perspective. Remember—the CAs are not the worker nodes; the PSEs are.

Let's now look at how an enterprise administrator logs in to the Zscaler administration portal and configures the policy that will be used by CAs to communicate down to PSEs for enforcement.

Admin Portal

The **Admin Portal** is a one-stop shop to manage the entire Zscaler configuration for an enterprise. Let's now examine the various aspects of configuration, after initial cloud instantiation, for an enterprise. These include the initial login, creating administrator roles, creating administrator and auditor accounts for everyday use, and so on.

Initial login

Zscaler operates many clouds, among these `zscaler.net`, `zscalerone.net`, `zscalertwo.net`, `zscalerthree.net`, and `zscloud.net`. The Admin Portal URL has admin in front of the cloud name—for example, `https://admin.zscalerthree.net`. The first time, log in using the super admin credentials that are sent to the main contact on the contract.

Upon the initial super admin login, it is strongly recommended to change that password with the proper complexity and lock the password away in a **multi-factor authentication (MFA)**-enabled secure location. Do not use the super admin account (also called default admin by Zscaler) to make changes. Changes performed by the `admin@companydomain.com` default admin account will have no accountability if something goes wrong.

Creating administrator roles

The next step would be to create administrator accounts. After logging in to the portal using the super admin password, click on the **Administration** icon on the left and, under **Authentication -> Administration Controls**, choose the **Role Management** option. Create roles such as administrator, helpdesk, auditor, and so on, and grant those roles appropriate permissions as per your company policy. Remember to activate your changes every time after you make a change by clicking on the **Activation** tab on the left, and click the blue **Activate** button. This makes sure changes are applied.

Creating administrator accounts

After the roles have been created, select the **Administration** icon on the left and, under **Authentication -> Administration Controls**, select **Administrator Management**. The main contact with Super Admin privileges can start by creating their account first. Then, they can add administrators, using the appropriate roles previously created with strong passwords, and have them log in to the portal, change their password, and verify their permissions. They can click on the **Administration** tab and then navigate to **Settings -> Account Management**, click on **My Profile**, and then change their password, language, and policy display preferences, as well as the time zone.

When a new administrator account is created, there is an option to enable Security, Service, and Product updates from Zscaler. **Security Updates** provide the latest information on threats and vulnerabilities that may affect an enterprise. **Service Updates** provide the latest information on new product and service enhancements, such as cloud updates and new data center additions. **Product Updates** provide communication about important Zscaler service changes and updates.

Administrator Management

Under **Administrator Management**, click the **Administrator Management** tab. You can choose to *disable* or *enable* password expiration. If you *disable* password expiration, the password never expires. If you *enable* password expiration, you will be asked to enter the number of days after which the password expires, and the administrator will then have to change their password. You can also enable **Security Assertion Markup Language** (**SAML**) authentication, which is out of the scope of this book, and hence a link has been provided at the end of this chapter.

Auditor account

The Zscaler Admin Portal has an inbuilt option to create an auditor account. While in the **Administrator Management** menu, click the **Auditors** tab. Click on **Add Auditor** and then enter the login ID, name, and password. Note that there is no option to select any permissions as this is automatically built in by Zscaler.

Usually, when an administrator with a role that obfuscates usernames logs in to the Admin Portal, they cannot see the actual usernames. In that case, an auditor can click the **Override** link on the top right and then enter the auditor credentials. The usernames will then appear normally.

Audit Logs

Audit Logs, as with any system, shows a list of which admins logged in from where, and when. **Audit Logs** also stores a log of all changes that were made by administrators. The administrator change logs are stored for up to 6 months by Zscaler and can be downloaded for a period covering 31 days at a time.

While logged in to the Admin Portal, click on the **Administration** tab on the left, and then navigate to **Authentication -> Administration Controls** and click on **Audit Logs**. Select the time range as needed, and then use one or more filters that are offered. These are **Action**, **Category**, **Sub-category**, **Interface**, and **Result**, outlined in more detail here:

- **Action** —This could be one or more of **Activate, Audit Operation, Create, Delete**, and so on.

- **Category**—The choices are **Activation, Administrator Management, Advanced Settings**, and so on.

- **Sub-category**—The options are **Account Info, Activation, Advanced Settings**, and so on.

- **Interface**—This could be **Admin User Interface, API, All**, and so on.

- **Result** —This could be **Failure, Success**, or **All**.

Finally, the logs can be downloaded by clicking on the **Download CSV** icon on the top right of the page.

Admin rank

The **admin rank** feature provided by Zscaler makes sure that changes made by an administrator with a higher admin rank cannot be overridden by an administrator with a lower admin rank. The admin rank ranges from **zero** (highest rank) to **seven** (lowest rank). For example, a security director has an admin rank of **zero**; a security manager has an admin rank of **one**; and a security engineer has an admin rank of **two**. Changes made by a security director cannot be overridden by either a security manager or a security engineer. Similarly, changes made by a security manager cannot be overridden by a security engineer but can be overridden by a security director.

To enable this option, click on the **Administration** tab and then navigate to **Settings** -> **Advanced Settings**. Under **Enable Admin Ranking**, click the tab to make it *green* and then save and activate your changes. Now, you can navigate to **Role Management**, create a new role or edit an existing role, and you will see the **Admin Rank** option.

> **Important note**
>
> Use this option with caution. In the preceding example, if a security director or security manager was not available and there was an urgent need to override a change to resolve an incident, then a security engineer would be unable to help.

Setting the company profile

The company profile can be set by the main contact on the contract (Super Admin) or an administrator who has been given the proper authority by that Super Admin. While logged in to the portal, click the **Administrator** tab on the left, then navigate to **Settings** -> **Account Management** and click on **Company Profile**. Verify that the company name and the associated domain(s) are provisioned correctly. Make a note of your company ID.

Fill in housekeeping information such as the street address and time zone of your company (headquarters, usually). Optionally, you can also upload a logo of your company. This will be used in messages displayed to your end users when the solution is deployed so that they are not confused by seeing a message without an official company logo.

Finally, fill in the two technical, billing, and business contacts. Zscaler sends out communications from time to time that will be received by these contacts.

Backup & Restore

As the Zscaler solution is deployed and evolves, there may be many changes that are made by many administrators over time. Hence, it is a good idea to create *restore points* that save a snapshot of the configuration. Click on the **Administration** tab, then navigate to **Authentication -> Administration Controls** and click on **Backup & Restore**.

Click on **Add Restore Point** and give the restore point an appropriate name, including the date, and add a description. Once the restore point is created, click on the *eye* icon on the far right to view the restore point. You have two options here. One is **View Stored Policies** and the other is **Restore Policy**. The first one allows you to see the policies that were captured by the restore point, and the second option allows you to overwrite the current policies with the policies in the restore point. If you decide to do the latter, it is strongly recommended to create a backup of the current configuration (possibly by creating another restore point).

In this section, we learned about the various configuration elements that are configured by an enterprise administrator via the Admin Portal, which are then stored by the CA. In the next section, we will examine the next component of the Zscaler cloud: the PSE.

Using Zscaler PSEs – where the policies are applied and enforced

The next important component of the Zscaler cloud is the PSE. Recall that Zscaler sits between the end user and the web destination; so, when the end user is trying to go out to the internet, their first stop is the PSE. The user web traffic directly hits the nearest PSE or the PSE configured by the company administrator.

The PSE being in the data plane, its task is to perform high-speed data-packet inspection and company policy enforcement. When the PSE encounters a new packet for which it does not know the company or user details, it performs a lookup to the CA and asks the CA for details about that packet. It extracts that randomized identifier we talked about in the previous section and uses it to query the CA. The CA returns the identity and the policy information for that identifier in less than a second.

For all subsequent data packets that are part of this traffic session, the PSE remembers this user. It caches this information so that it does not have to query the CA again and again for the same situation. This allows the subsequent data-packet processing to be extremely fast and efficient.

The speed also comes from the fact that the data-packet processing is performed in memory only (somewhat equivalent to a **Random-Access Memory** (**RAM**) in a personal computer). The data is never written to disk during this processing phase.

Let's now look at what **Secure Sockets Layer** (**SSL**) inspection is and why it is so important for enterprises today in the context of web traffic security.

SSL inspection

Gartner Research (`https://www.extrahop.com/company/blog/2020/gartner-report-on-handling-challenges-with-tls-1.3/`) estimates that about 80% of enterprise web traffic was SSL-encrypted in 2019 and that the percentage continues to increase every day. Gartner also predicts that over 70% of attacks by bad actors in 2020 will use some type of encryption. At the time of this writing, according to *Google Transparency Report* (`https://transparencyreport.google.com/https/overview?hl=en`), 95% of the traffic seen across Google was encrypted.

So, why is SSL inspection important? Because many enterprises think that encrypted traffic is safe. However, the problem is that bad actors are using encryption for their attacks, hence SSL inspection is extremely important for enterprises. If enterprises do not enable SSL inspection, then they are unaware of over 80% of traffic going into and out of their environment. This is a huge, gaping hole in their security posture.

This is the primary reason that SSL inspection is highly recommended for enterprises. Another reason is to gain visibility into user traffic activity, and it allows Zscaler to apply policies granularly down to the user, location, group, or department level.

With Zscaler SSL inspection, enterprises can use the certificate that Zscaler provides by default, or the enterprise can generate a custom certificate (signed by their own internal **public key infrastructure** (**PKI**) team) if their security posture requires it. If a Zscaler certificate is used, then the enterprise infrastructure team will need to distribute the Zscaler root CA (here, **CA** means the **Certificate Authority**, not Zscaler Central Authority) certificate to all end users so that they can trust the connection to Zscaler.

If the end users or the enterprise is concerned about their personal data privacy, exemptions can be configured to certain types of web destinations. Usually, financial and/or healthcare website destinations are added to this exemption list. More information on this exemption will be provided in later chapters.

SSL inspection flow

Here are the steps that occur when an end user tries to navigate to a destination URL:

1. The end user initiates an outbound SSL connection to a destination server. As Zscaler is the proxy between the end user and the destination URL, the end user's outbound SSL request is handled by the PSE.

2. The PSE will in turn initiate another SSL connection to the destination URL.

3. The PSE receives a reply from the destination server that has the server's certificate.

4. The PSE will now reply to the originating end user with a certificate signed using a key managed by Zscaler.

5. Since the end-user device already has the Zscaler root certificate, it uses its public key to validate the certificate received on this new connection. If, instead, a custom certificate was already uploaded to Zscaler, the end-user device should already have the root certification from the enterprise's CA.

6. An SSL connection is then established between the end user and the PSE.

7. The PSE, in turn, establishes a connection with the destination server of the URL.

In summary, the PSE is now in between the end user and the destination server of the URL and can inspect the traffic and apply the proper policies that it gets from the Zscaler CA.

Virtual Service Edge (formerly called Virtual ZEN or VZEN)

We just saw that end users usually access the destination web servers through the PSE nodes. This means that from the destination web server's point of view, the end user is coming from a Zscaler **Internet Protocol** (**IP**) address and not the enterprise's IP addressing scheme. This may not be acceptable to some companies who rely on using their enterprise IP-addressing scheme for various reasons.

Enterprises can subscribe to a product called a **Virtual Service Edge** (**VSE**) at an additional subscription fee from Zscaler. A VSE is a VM that can be called a **PSE in a pocket**. A VSE performs the same function as an actual PSE in the Zscaler cloud. It works with the Zscaler cloud, downloads the enterprise's configured policies and settings, and performs enforcement on the enterprise's traffic.

The use cases for VSEs are outlined here:

*   Geopolitical requirements, restrictions, and regulations sometimes allow access to certain websites only from certain locations.

*   Zscaler PSEs are not available in all countries, so enterprises may sometimes have to connect to a PSE in a different country.

*   Some enterprises use their IP address range as one of the authentication factors when accessing certain vendors. Those vendors used to the enterprise's IP address range may not be open to the idea of whitelisting Zscaler IP address ranges.

*   When PSEs are not located in proximity, this can cause excessive, intolerable delay for an enterprise.

Let's now explore the architecture and requirements for a VSE so that an enterprise can make a decision whether to deploy a VSE or not.

VSE architecture and requirements

As mentioned earlier, a VSE is a VM that is deployed on the customer internal network and uses the enterprise public IP for outbound web traffic. VSEs are typically deployed in pairs (although they can work in a standalone mode), keeping in mind the principle of high availability. The enterprise sends its traffic to the VSE just as if it were sending the same traffic to an actual PSE. The VSE only needs an outbound connection to the Zscaler cloud, and an inbound connection is only needed when the enterprise wants a Zscaler associate to assist in troubleshooting the VSE. Up to 20 VSE nodes can be deployed in a cluster for additional performance.

A single VSE supports up to 600 **megabits per second** (**Mbps**) traffic throughput. An SSL acceleration card is recommended anytime the throughput needs to be at least 100 Mbps. The host requirement for the VSE is a VMWare ESX/ESXi hypervisor. The hardware specifications, setup, install, and configuration instructions for the VM can be obtained from Zscaler's documentation. This is quite an involved process and is out of the scope of this book.

A VSE cluster typically consists of at least two VMs, with an internal load balancer directing traffic to both nodes. When one node goes down, the other continues to forward traffic. Zscaler updates the software on the VSEs on a regular basis. For zero downtime, Zscaler upgrades the software on one VSE node at a time in the cluster. When that upgrade is done and verified, Zscaler updates the second VSE. VSE health and status can be monitored by using **Simple Network Management Protocol** (**SNMP**) monitoring tools to take proactive action in the case of a hardware failure.

To verify if an enterprise has a VSE subscription, log in to the Admin Portal and click on the **Administration** tab. Navigate to **Settings -> Account Management** and click on **Company Profile**. Click on the **SUBSCRIPTIONS** tab and look for the **Virtual Service Edge** component, and note the status and the service start and end dates. If you do not see the item in the list at all, then your enterprise has not purchased this subscription.

To download the **Open Virtualization Appliance** (**OVA**) file for the VSE, click on the **Administration** tab. Navigate to **Settings -> Cloud Configuration** and click on **Virtual Service Edges**. In the first tab, **VIRTUAL SERVICE EDGES**, click on **Download Virtual Service Edge VM**.

In this section, we understood the functions of the Zscaler's worker nodes—the PSEs—and how they work in conjunction with the CA to perform high-speed packet processing including SSL inspection. We also saw how an enterprise can implement a "pocket ZEN" in their own network based on the business use case. In the next section, we will learn about the third component of the Zscaler cloud: the Nanolog cluster.

Monitoring user and application activity using Nanolog clusters

The third component of the Zscaler cloud is the **Nanolog**, which deals with analytics and reporting. This component is important from both a customer and Zscaler perspective. The customer would like to know what kind of data is flowing through their enterprise systems, such as which internet applications are being used, which locations are using the most bandwidth, and if that bandwidth is being spent on productive applications supporting the company's bottom line.

From a Zscaler perspective, it gives information about where the threats are coming from and what types of threats are evolving, and helps Zscaler plan for future growth by adding more capacity to its cloud locations. Of course, it can also be used for advertising and marketing purposes.

Each request from a user appears to be a simple request for a web page on the internet, but the dynamic nature of today's web pages means that this does not just include text, but also images, graphics, embedded videos, and so on. This means that a single request from a user now consists of multiple sub-requests, each loading a little object that makes up the main request.

This means that the number of logs generated by each user accessing the internet is a high one. Multiply that by the total number of users an enterprise has, and this number quickly reaches the hundreds of thousands within minutes. Also, when users travel to a different location from where they were originally provisioned, that must be correlated as well across different geographies.

Simply storing or streaming all these logs in their raw form is a difficult, non-scalable, and expensive proposition. Instead, the Nanolog infrastructure applies structuring to these logs by using optimization and compression in real time as these logs are generated, and sends them to a component called Log routers.

The **Log routers** component uses an intelligence-based mechanism, such as where the user is based and where the company wants to store its logs (such as **General Data Protection Regulation** (**GDPR**) requirements), and then correlates this information, writing the logs to disk in the geography of the customer's choice. This has become increasingly necessary for enterprises based on their data security and data regulatory requirements.

For example, a company based in the **European Union** (**EU**) probably cannot have—and does not want—its logs to be written to disk and stored in the Americas or the Asia-Pacific region. All the Nanolog infrastructure does is send the logs to Log routers, which correlate the requirements and redirect the logs to be written as dictated by the customer's needs. It does not matter if that company's employees are traveling to a different region, as those employees' logs are tied back to the company's physical location.

Nanolog Streaming Service (NSS)

NSS is a VM running inside the enterprise's network or **demilitarized zone** (**DMZ**) and not just limited to an on-premises installation. The NSS sits between the customer's SIEM and Zscaler's Nanolog infrastructure. Zscaler's Nanolog infrastructure streams the logs to the NSS using a secure, certificate-based **Transport Layer Security** (**TLS**) connection. The NSS will in turn decompress the stream and send it to the customer's SIEM such as Splunk, ArcSight, or IBM QRadar, or even to Syslog in a **Comma-Separated Value** (**CSV**) format. The actual setup and configuration of NSS is out of the scope of this book.

Here are some of the benefits of using NSS for an enterprise:

- Zscaler Nanolog servers store logs for 6 months. If an enterprise needs to store logs for a longer duration, they can use NSS.

- The Zscaler logs can be streamed by the NSS to a SIEM of the customer's choice. Those logs can then be correlated with logs from other devices in the enterprise such as routers, switches, firewalls, active directory servers, and so on.

- NSS offers multiple customized streams and flexible log formatting and filtering.

- The usernames can be obfuscated by the enterprise to comply with **human resources** (**HR**) or industry regulations.

- The software on the NSS VMs is automatically kept up to date by Zscaler.

Let's now explore the NSS VM architecture, the supported platforms and requirements, and configuration steps for feeds and alerts.

NSS VM architecture

NSS VMs are usually deployed in pairs, one for the web logs and one for the firewall logs. Zscaler recommends that these NSS VM pairs be deployed in a redundant configuration—that is, deploy two NSS VMs for web logs and two for firewall logs (note that one VM each for web logs and firewall logs can still work). That way, in case one NSS VM goes down, the other continues to stream the logs.

If the NSS VM ever loses its connection to the customer SIEM, the NSS VM will automatically continue to buffer the logs until the connection is restored. Depending on the number of logs flowing through the system, additional RAM may need to be added to the VM for sufficient buffering capacity. Using the minimum RAM recommendation will allow buffering of up to 1 hour's worth of logs.

When connectivity is restored, NSS can be configured to send logs, starting a few minutes before the time at which the connection was lost. Each log is identified by a unique record ID so that the SIEM can eliminate the duplicate logs. The same behavior applies to the connectivity loss between the NSS and the Nanolog infrastructure.

NSS VM supported platforms and requirement calculations

NSS VMs can be provisioned on traditional on-premises hypervisors such as VMWare ESX/ESXi or on public cloud platforms such as AWS or Azure. The requirements and configuration steps for NSS VMs can be obtained from Zscaler's documentation. This is quite an involved process and is out of the scope of this book.

To estimate the requirements for your enterprise, log in to the Admin Portal and click on the **Administration** tab. Navigate to **Settings -> Cloud Configuration** and click on **Nanolog Streaming Service**. On the **NSS SERVERS** tab, click on the **Deploy NSS Virtual Appliance** link. Select your NSS Type, enter the number of users in your enterprise, and an estimated peak transactions per hour. Select your platform—namely, VMWare, AWS, or Azure—and click on **Compute**. Note down the recommended VM and hypervisor specifications and download the NSS Virtual Appliance as a .ova file.

NSS feeds and alerts

Feeds can be configured on the NSS that detail where and how logs are sent (to the SIEM). The steps involved in creating a feed are as follows:

1. After navigating to the **Nanolog Streaming Service** section, as mentioned previously, click on the **NSS FEEDS** tab.

2. Click on the **Add NSS Feeds** option. Enter basic details such as **Feed Name, NSS Server, SIEM IP address and SIEM TCP port, SIEM Rate, Log Type, Feed Output Type, User Obfuscation, Time zone, Duplicate Logs,** and **Log Filter** settings.

3. Activate your changes, and then wait a few minutes for the logs to start streaming. Then, log in to your SIEM and verify that you are receiving the logs.

Setting the **SIEM Rate** to **Limited** would then throttle the events per second that are streamed to your SIEM solution. A setting of **Unlimited** would not place any throttling restrictions. The **Log Type** can be **Web Log, Tunnel,** or **Alert**. **Feed Output Type** can be set to **CSV** or be a different option, based on the vendor make and model of the SIEM. **User Obfuscation** can be set to **Enabled** or **Disabled** (we discussed obfuscation earlier). The **Duplicate Logs** setting controls how far back the NSS should resend the logs after detecting a loss of connectivity between itself and the SIEM.

In this section, we saw how Nanolog clusters handle the problem of ever-increasing velocity of web transaction logs and intelligently route them to Nanolog routers, and how they are written to disk in a geography of the customer's choice. We also saw how NSS could be used by an enterprise to store logs for a longer duration and send the logs to their SIEM for intelligent and advanced correlation. In the next section, we will examine the last component of the Zscaler cloud: Sandbox.

Protecting enterprise users and infrastructure with Sandbox

When a malicious piece of software is identified, its signature is usually added to a database. The next time the same signature is observed, it can quickly be flagged after comparing it with a signature already in the database. But how can we identify the signature the first time?

Understanding the need for a sandbox

Enter the sandbox as the solution. In the cloud, it is very easy to provision a sandbox environment, often using automated scripts, then the new piece of code that needs to be tested is executed (a.k.a. detonated) in that environment. From the resulting data, attention is focused on the adverse effects of that code on the environment. Once this is recorded, it is added to the database and propagated to the entire cloud very quickly.

Why do we prefer a cloud sandbox? Consider doing this in an on-premises solution. You need to first provision the hardware necessary and then set up your environment such as the operating system, and so on. This may take several days, or weeks, to set up. Once you find a malicious signature, you will then have to record it, copy it to disk, and then distribute it to the other parties who have subscribed to your service. This whole end-to-end process does not happen in real time. Then, you must keep refreshing your hardware on a periodic basis.

Now, consider the Zscaler cloud sandbox. It takes very little time (usually minutes) to spin up a VM with the necessary operating system configuration using **application programming interfaces (APIs)**, after which you can detonate your piece of code and then record its effects. Now, propagate this information to all the clouds within minutes. Since Zscaler analyzes its packets in the cloud, the PSEs can get these signature updates in real time and act on the packets right away. There is no need to wait until the customer updates their threat signatures, and then they take effect. As far as the hardware refresh cycle goes, you can simply add generic hardware to the sandbox clusters and carve out your VMs as needed.

Here are the salient features of the Zscaler Cloud Sandbox:

- It sits in-line with the traffic between the end users and the internet, so nothing escapes its view.

- It has access to the threat intelligence database used to identify already identified malicious signatures and detonates any unknown signatures in its environment, as explained earlier.

- As soon as a new malicious signature is detected, it immediately starts blocking that malicious signature while updating the threat intelligence database.

- It can scan and inspect all types of files, even the ones that are SSL-encrypted.

Cloud Sandbox collects all the suspicious payloads from all the PSEs and forwards them to the sandbox clusters. In those centralized clusters, the unknown payloads are detonated in a safe, controlled environment. If an unknown payload is found to be malicious, the sandbox immediately informs the PSEs of the outcome so that the PSEs can start implementing the protection mechanism against that malicious signature.

Zscaler's standard Cloud Sandbox is an included feature with ZIA. It allows Windows executable (.exe) and Windows library (.dll) file types to be inspected, and the file size is limited to 2 **megabytes (MB)**. If an enterprise deals with larger than 2 MB files on a frequent basis and does not already have a sandbox product, they can subscribe to the advanced sandbox option for an additional fee.

Zscaler's advanced sandbox feature can scan many additional types of file such as `.doc(x)`, `.xls(x)`, `.ppt(x)`, `.pdf`, and so on. There is no restriction on file size, and additional options are available such as file quarantine, granular policy controls, and extensive reporting.

Cloud Sandbox configuration options

The overall action that Cloud Sandbox takes depends on two fields—**First-Time Action** and **Action for Subsequent Downloads**. Here are the options for **First-Time Action**:

- **Allow and do not scan**: This option means that you allow the download to go through and do not send the file to sandbox scanning.

- **Allow and scan**: This option means that you allow the download to go through and send the file to sandbox scanning.

- **Quarantine**: This option means that you hold the file in quarantine, display a wait message to the end user that the file is being scanned, and let the sandbox complete its scanning. If the scan is clean, you release the file to the end user. If the file is found to be malicious, display a message to the end user that the file was blocked.

There are two options for **Action for Subsequent Downloads—Allow** and **Block**. This field defines the behavior to follow when the same file is downloaded again. We can either allow or block this action.

Let's now look at some real-world examples of combinations of these two fields, as follows:

- **Quarantine First Time, Block Subsequent Downloads**: The downloaded file is not available until Cloud Sandbox tests it and gives it an all-OK and releases it to the end user. This is the most conservative option when it comes to security.

- **Allow and Scan First Time, Block Subsequent Downloads**: This is a middle-ground option whereby the file is immediately released to the end user but is also sent to scanning in the background. If the file turns out to have a malicious signature, the damage is probably already done by then.

Cloud Sandbox can be configured based on business needs. Some enterprises can tolerate the delay from the time an end user requests a file download till when Cloud Sandbox determines the file is clean (the first option listed previously). This is very similar to an anti-virus software scanning a file before letting you open it. Some enterprises cannot tolerate this delay and want the file available immediately, and then additional scanning can take place in the background (the second option listed previously). Each option has its trade-offs.

During the download process, the end user is shown appropriate messages so that they are informed of the sandbox policy their organization has chosen to configure. With the first option listed previously, they are shown a message that their file is being scanned and, if found to be safe, it will be released to them. After scanning is done, if the file is found to be malicious, they are then shown a block notice. There are no messages shown for the second option as the file download is allowed the first time.

If a sandbox rule is configured with the first-time action of **Allow and do not scan** and **Allow for all subsequent downloads**, no sandbox scanning will take place at all. It is hard to understand why anyone would use such a combination and purposefully bypass sandboxing, and hence we did not mention it in the real-world examples shown previously.

Multiple sandbox rules can be created and those rules are generally numbered, with the lowest number first. It is generally recommended that the rule addressing the most common file options is included at the top of the list, followed by sandbox rules that address the less common file types. At the end of the list is the Zscaler's default sandbox rule that cannot be deleted but can only be edited. Each sandbox rule also has a rich set of fields that can be used, such as file types, URL categories, users, groups, departments, locations, sandbox categories, and protocols. A detailed discussion of all options is beyond the scope of this book.

Zscaler's default option blocks all Windows executable (`.exe`) and Windows library (`.dll`) files from suspicious URLs. If an end user downloads such a file, Zscaler by default releases the file to the end users while sending the file to the sandbox cluster for scanning. It is to be noted here that due diligence is completely the responsibility of the enterprise; the enterprise cannot outsource that decision to Zscaler. If they do, this means that they are accepting Zscaler's default options. By the same token, this book does not make any recommendations either and leaves the choice to the enterprise itself.

With the **Advanced Sandbox** option, the enterprise gets a wealth of information on sandbox reporting. The detailed sandbox report shows the following:

- Classification of the malware and its threat score (a score of less than 40 is benign; a score between 40 and 70 is suspicious, but not blocked; a score of 70 and above is considered malicious and is blocked)
- The behavior, symptoms, and screenshots of the malware
- How the malware compromises or exploits the system
- The kind of network activity that the malware does

- How the malware tries to persist within a system and how it spreads itself
- A list of all files that were quarantined
- Where the malware originated from and the attributes of its signature

Armed with this type of information, the security team of an enterprise can closely work with the information technology and infrastructure team to educate the end users. Based on these reports, they can also identify potentially infected computers and isolate and clean them on a proactive basis, thereby strengthening the security posture for the entire enterprise.

In this section, we learned about the need for a sandbox and how a cloud sandbox is necessary for the internet of today. We also saw the advanced sandbox feature of Zscaler that offers more functionality and an expanded feature set over standard sandbox.

Summary

In summary, we saw that the three major components of the Zscaler cloud—namely, CA, PSE, and Nanolog—are architected in a decoupled manner. This means that each component can be scaled up or down based on its dynamic needs, and the other components will not miss a beat. Each component is also designed to be highly available at every level (component level, data center level, and so on).

It is very important to understand the planes in which each of these components operate and what their individual role is in the entire end-to-end transaction that occurs between the end user and the destination website or web application. This understanding is critical to avoid frequent confusion, moving forward.

We also saw how SSL inspection works and why it is important as more and more web traffic (and bad actors!) continue to use encryption. The use cases for VSEs and sandbox and its various configuration options to suit enterprise needs were also covered. In the next chapter, we will get into learning about ZIA features in depth.

Questions

As we conclude, here is a list of questions for you to test your knowledge regarding this chapter's material. You will find the answers in the *Assessments* section of the *Appendix*:

1. What are the four components of the Zscaler cloud?

2. Where is the configuration for the enterprise stored in the Zscaler cloud?

    a. CA

    b. PSE

    c. Nanolog cluster

    d. Sandbox

3. The PSE processes the packet in disk memory using a service chain.

    a. True

    b. False

4. SSL inspection is not considered necessary as part of a strong security posture.

    a. True

    b. False

5. By default, the Nanolog clusters store logs for up to:

    a. 3 months

    b. 6 months

    c. 9 months

    d. 12 months

6. A sandbox detonates an unknown piece of code in a safe environment.

    a. True

    b. False

Further reading

- Configuring SAML for ZIA: `https://help.zscaler.com/zia/configuring-saml`

3
Delving into ZIA Policy Features

In this chapter, we will learn about the various features, traffic forwarding methods, authentication mechanisms, and the **Zscaler Client Connector** (**ZCC**).We will learn what each feature does and what the industry best practice is for using that particular feature. Luckily, Zscaler enables most of the industry-standard practices by default. After this chapter, we will be ready to start planning a customized **Zscaler Internet Access** (**ZIA**) solution for a small business or a large enterprise.

ZIA policy features are the most important elements and can be categorized into three areas–Web, Mobile, and Firewall. The Web policy is extensive and mainly consists of security, access control, and data loss prevention. The Mobile section consists of ZCC configuration, security, and access control. The Firewall policy offers firewall control, **network address translation** (**NAT**) control, **Domain Name System** (**DNS**) control, **File Transfer Protocol** (**FTP**) control, and **intrusion protection system** (**IPS**) control. These sections will be explained in the same order as they appear in the policy section of the Zscaler ZIA Admin Portal to make it easy to follow.

Finally, we will end with the enforcement order of these various policies so that an administrator can visualize the traffic flow.

In this chapter, we are going to cover the following main topics:

- Understanding the ZIA Web policy
- Exploring the ZIA Mobile policy
- Learning about the ZIA Firewall policy
- Order of policy enforcement

Technical requirements

A general understanding of the various web security terms and technologies will aid in understanding the content of this chapter. Access to the Zscaler Admin Portal will make it easier to follow along when discussing each policy feature and its location in the Portal.

Understanding the ZIA Web policy

The core function of ZIA is to protect enterprise end-user web traffic from threats that are associated with accessing the internet and internet applications. That is why we need to understand all the secure web policies offered by ZIA.

Administrators can click on the **Policy** tab on the left after logging in to the Zscaler Admin Portal to access these options. On each page, there is a **Recommended Policy** link in the upper-right corner. If a company administrator is unsure of the industry-standard best practice, they can click on this link and view the Zscaler recommended policy.

The ZIA Web policy can be categorized into three sub-sections—**Security**, **Access Control**, and **DLP**. The security policy features control the most common aspects of web traffic, such as what type of traffic to inspect, the behavior for each type of malicious traffic, and any security exceptions. The **Access-Control** component controls end-user access to each type of web traffic, such as streaming, productivity, and social media. Finally, the DLP component helps enterprises to translate their data protection needs into actual DLP policies. Let's examine in detail each section and the features offered within each section.

Security

Security is the first category under **Web**, which offers policies for the most common threats seen when using the internet. It consists of four main components—**Malware Protection**, **Advanced Threat Protection**, **Sandbox**, and **Browser Control**. Let's explore these components in detail.

Malware Protection

One of the most common attack mechanisms employed by bad actors is to inject malware into enterprise systems via the internet, and then either compromise the integrity of or steal data. The **Malware Protection** page has two tabs—**Malware Policy** and **Security Exceptions**. The sections detailed next are included in the **Malware Policy** page.

Traffic Inspection

There are two options available for traffic inspection. We can configure inspection for either **inbound traffic** or **outbound traffic**. Inbound traffic is traffic toward an enterprise, and outbound traffic is traffic toward the internet. This allows traffic to be analyzed in real time for the protocols selected by an enterprise. Files with up to five layers of recursive compression are supported for inspection.

Protocol Inspection

There are three choices available under this section. The first one is to inspect **HTTP (and HTTPS if SSL inspection is enabled)** traffic depending on the direction of traffic configured under **Traffic Inspection**, discussed previously. The second one is to inspect **FTP Traffic** running over the **HyperText Transfer Protocol (HTTP)**. The third—and last—one is to inspect **Native FTP Traffic**.

Malware Protection

There are several options under this section. The options are **Viruses**, **Unwanted Applications**, **Trojans**, **Worms**, and **Ransomware**. All these options have either an **Allow** or a **Block** option.

Adware/Spyware Protection

The options available here are **Adware** and **Spyware**. As with the preceding option settings, you can set these to either **Allow** or **Block**.

The Zscaler-recommended policy for **Traffic Inspection** and **Protocol Inspection** is to enable all options. For malware, adware, and spyware protection options, the recommended policy is to block them all. Enterprises using the ZIA service are ultimately responsible for the options they choose as part of their deployment. Let's now see what is included in the **Security Exceptions** tab.

Security Exceptions

Enterprises can set a few security exceptions here. They can either allow or block password-protected files and unscannable files. These options depend on the business needs and requirements of the enterprise. If scanning is not needed for content from certain **Uniform Resource Locators** (**URLs**)—such as trusted websites—they can be configured here as a list, with an upper limit of 25,000 URLs.

Advanced Threat Protection

The **Advanced Threat Protection** page consists of two main tabs—**Advanced Threats Policy** and **Security Exceptions**. Let's now learn about the components of the **Advanced Threat Policy** tab.

Suspicious Content Protection

Zscaler checks each web page for any vulnerabilities and harmful content and assigns a risk rating for those pages. These risk levels are categorized as low (about 0 to 33), moderate (about 34 to 66) and high (about 67 to 100). As the risk value increases, so does the probability that the web page is more and more suspicious and/or malicious. The default risk value that Zscaler configures is 33.

> **Important note**
> It is up to the enterprise using this product to decide on the value that is right for them. Enterprise administrators can get the risk rating for a web page using **Zscaler Zulu URL Risk Analyzer** and submit the URL in question, and then wait for the rating value to be calculated by the tool and displayed. There is a risk of accidental blocking if the risk rating is set too low.

Botnet Protection

The options under this section are **Command and Control Servers** and **Command and Control Traffic**. The first option checks against known **command-and-control** (**C&C**) servers, and the second option checks for traffic sending and receiving commands from unknown servers. The option settings are **Allow** or **Block**.

Malicious Active Content Protection

The available options here are **Malicious Content & Sites**, **Vulnerable ActiveX Controls**, **Browser Exploits**, **File Format Vulnerabilities**, and **Blocked Malicious URLs**. The option settings for the first four options are **Allow** or **Block**. The fifth option gives enterprises the ability to configure known malicious URLs as a list, up to a maximum limit of 25,000 URLs.

Fraud Protection

The various options under this section are **Known Phishing Sites, Suspected Phishing Sites, Spyware Callback, Web Spam, Crypto Mining**, and **Known Adware & Spyware Sites**. The option settings are **Allow** or **Block**.

Unauthorized Communication Protection

These options protect enterprises against communication channels not authorized by the enterprise. The options are **IRC Tunneling** (**IRC** stands for **Internet Relay Chat**), **SSH Tunneling**, and **Anonymizers**. The option settings are **Allow** or **Block**.

Cross-site Scripting (XSS) Protection

The options available here are **Cookie Stealing** and **Potentially Malicious Requests**. The option settings are **Allow** or **Block**.

Suspicious Destinations Protection

If an enterprise has no business connections with certain countries or if they are prohibited by their governments from conducting transactions with certain countries, this option can be used. Under **Blocked Countries**, you can select the names of the countries and the enterprise users will be blocked from content from those countries.

P2P File Sharing, Anonymizer, and VoIP Protections

In **Point to Point** (**P2P**), there is no intermediate server between two end users or endpoints, and communication is directly between two endpoints or end users. Under **P2P File Sharing Protection**, you can **Allow** or **Block** the BitTorrent file-sharing application. Under **P2P Anonymizer Protection**, you can block Tor, which is an anonymized protocol that deals only with encrypted data and hence cannot be inspected. Under **P2P VoIP Protection**, you can **Allow** or **Block** the Google Hangouts application.

Let's now examine the features under the **Security Exceptions** tab of the **Advanced Threat Protection** page.

Security Exceptions

Under this tab, you can add a list of URLs (up to 25,000) under the **Do Not Scan Content from these URLs** option, and ZIA will not scan for malicious content from these URLs. It would be too lengthy to list the Zscaler-recommended policy settings for **Advanced Threat Protection** here. The recommended policy can be obtained by clicking on the **Recommended Policy** link on the top-right side of the page.

Sandbox

The Zscaler Admin Portal already has a default sandbox rule configured that can only be edited and cannot be deleted. This default sandbox rule allows and scans the first time, and blocks subsequent downloads for the Microsoft Windows library and executable files from suspicious destinations. Other available options were already discussed in detail in *Chapter 2, Understanding the Modular Zscaler Architecture*.

Browser Control

Vulnerabilities are being discovered almost daily with many popular internet browsers. Many enterprises (such as small or medium businesses) do not have a dedicated IT team and hence cannot keep track of the browser versions their employees are using and the patching that is needed on a regular basis. This section offers them some options.

Browser Vulnerability Protection

By default, the **Enabled Checks & User Notification** option is turned off. When this option is turned on, enterprise administrators can set a few options. The **How Often to Check** option can be set to **Daily**, **Weekly**, or **Monthly**. If set to **Daily**, ZIA will check the end user's browser make and version every day. There is an option to enable or disable warning notifications for browsers. There are also two more options to disable notifications for plugins or applications.

Browser Blocking

If an enterprise chooses to allow only certain browser versions and block lower versions (instead of just warning the users), they need to disable the **Allow All Browsers** option. This adds more choices for the most popular browsers—namely, **Microsoft Browsers**, **Chrome**, **Firefox**, **Safari**, and **Opera**. Enterprises can choose specific options under each browser. Make sure end users are aware of the browser versions approved by the enterprise so that they know which browser and which version to get on their computer.

Note that the **Sandbox** and **Browser Control** pages have no **Security Exceptions** tab, unlike the **Malware Protection** and **Advanced Threat Protection** pages.

We have explored the various available options for security and how they protect end users from threats on the internet. Now, let's move on to the second component of the ZIA policy—namely, **Access Control**.

Access Control

Access Control is the second major component under **Web**. **Access Control** enables enterprise administrators to define and control how end users can access internet applications and fine-tune that behavior. **Access Control** components are **URL & Cloud App Control**, **File Type Control**, **Bandwidth Control**, and **SSL Inspection**. Let's look at these in more detail.

URL and Cloud App Control

Both the **Cloud App Control** and **URL Filtering** policies offer **Block**, **Allow**, or **Allow with Caution** filtering rules. However, there is a major difference between **URL** and **Cloud App Control** filtering. Let's now see what that difference is.

It is laborious for an enterprise to keep track of all the URLs that form a part of a certain cloud application type. For example, if an enterprise needs to keep a collection of all email service providers such as Yahoo Mail, Gmail, and so on, this is a very tedious job to keep up with. New email service providers come up from time to time, and existing providers go out of existence.

For this reason, Zscaler offers broad collections of cloud application types, such as **Instant Messaging**, **Social Networking**, **Streaming Media**, and **Webmail**. These are the starting points for enterprises if they wish to modify these *collections*, saving a lot of time for enterprise administrators. These collections are called **Cloud Apps**.

Under **URL Filtering**, ZIA provides one default rule that blocks content from websites usually considered to add no value to an enterprise's employee productivity, such as sites related to gambling, nudity, and pornography. It is to be noted that all default **Cloud Apps** and **URL Filtering** rules are in a disabled state until they are enabled by the enterprise administrator.

By default, the **Cloud App Control** policy is enforced first, and then the **URL Filtering** policy. So, if the **Cloud App Control** policy blocks a certain website, the overall action by Zscaler is to block that website and not even consult the **URL Filtering** policy. Similarly, if the **Cloud App Control** policy allows a certain website, the overall action is to allow that website even if the **URL Filtering** policy blocks the same website. Zscaler strongly recommends the use of **Cloud App Control** over **URL Filtering**.

However, if the enterprise has a business case to use both options, there is a setting that can overwrite this default behavior. In the Admin Portal, under **Administration -> Advanced Settings**, enterprise administrators can enable the **Allow Cascading to URL Filtering** option. With this option enabled, even if the **Cloud App Control** policy allows a certain website and the **URL Filtering** policy blocks it, the overall action is to block that website.

URL Filtering and Cloud App Control rule options

There are several common choices offered when creating a new **URL Filtering** rule or a new **Cloud App Control** rule. Let's explore those options now:

- **Rule Order**—The rule order refers to where the rule appears in the entire list. It can be at the top of the list (first position) or it could be fourth on the list. Rules are evaluated in the order they are listed, and the first match is selected. It is generally a good idea to keep the most specific rules at the top and the more general rules further down.

- **Admin Rank**—We already talked about this option in *Chapter 2, Understanding the Modular Zscaler Architecture.*

- **Rule Name**—Give the rule a descriptive name so that it is easy to tell what the rule is supposed to do. That way, we can immediately check if the rules are properly configured.

- **Rule Status**—This can be enabled or disabled. During a ZIA deployment, it is common to configure certain test rules for specific test users. After troubleshooting, instead of deleting the rule and having to recreate it again, enterprise administrators can simply disable it.

- **Cloud Applications**—This allows the selection of one or more cloud applications under a specific category.

- **Users, Groups, Departments, Locations and Time**—After the user authentication is set up, rules can be created based on these criteria. The **Users, Groups,** and **Departments** fields are evaluated using a logical OR and the **Locations** and **Time** fields are evaluated with a logical AND. Every **Cloud App Control** rule or **URL Filtering** rule has all these options.

- **Action**—Under this setting, there are options relevant to the application itself. For example, for a **Webmail** application, the options are **Viewing and Sending Mail** and **Sending Attachments**. The option settings for these are **Allow** or **Block**. If the application is a web page itself, the option settings would be **Allow, Caution,** and **Block.**

- **Description**—This is a free-flowing text field that can be used to write a detailed description. This will also show up on the main **URL & Cloud App Control** page, so it is easy for an enterprise administrator to understand what this rule does, without the need to edit the rule and see what it does.

- **Unified Communications as a Service**—In this section, enterprises can choose to allow or block the communication platforms such as Zoom, LogMeIn, and RingCentral.

The **URL & Cloud App Control** page also has an **Advanced Policy Settings** tab. This tab offers a few advanced policy settings. Under **CIPA Compliance Report**, the **Enable CIPA Compliance** option can be enabled to enforce **CIPA (Children's Internet Protection Act)** compliance. It addresses concerns about children's access to objectionable content on the internet.

The other options that can be enabled are **Enable Newly Registered Domain Lookup**, **Enable AI/ML based Content Categorization**, **Enable Embedded Sites Categorization**, **Enforce SafeSearch**, and **Enable Identity-based Block Override**.

Under **Office 365 Configuration**, enterprise administrators can turn on the **Enable Microsoft-Recommended One Click Office 365 Configuration** option. Enabling this option relieves enterprises from constantly keeping track of all the Office 365 URLs and provides end users with a smooth Office 365 experience. There is also an option here to **Allow** or **Block** Skype access.

URL Categories

It is also relevant to bring up the **URL Categories** section here. Enterprise administrators can access this page by clicking on **Administration -> Resources -> Access Control -> URL Categories**. This page has the entire listing of the **URL Classes**, **Super-Categories**, and **Categories** that have been defined by Zscaler. These classes can be moved around and edited, and new categories can also be added.

For example, video-streaming applications are listed under the super-category named **Bandwidth Loss**. But if an enterprise is in the music industry, its employees need to watch videos as part of their job. Then, the **Video Streaming** category can be moved to the **Business Use** super-category, under **Corporate Marketing**.

If the listed URL categories do not apply to an enterprise, new URL categories can be added. By default, up to 64 custom categories can be created (this can be increased to 256 by requesting Zscaler support), and up to 25,000 custom URLs can be added across all custom URL categories and policy settings combined.

File Type control

This control is not the same as the **Sandbox** control. The **Sandbox** control provides the ability to scan certain types of files from certain types of websites and then allow the download to occur or block it. The **File Type** control provides the ability to impose a blanket ban for upload, download, or both on certain types of files for certain users and for certain websites, and so on.

For example, if an enterprise has kiosks that are used by customers to perform very restricted functions such as place an order or check in to a system, those terminals do not need the ability to download files as it has no use case. Such endpoints can be configured so that they are not able to download files.

Bandwidth Control

Many small- and medium-business enterprises struggle with limited bandwidth for their internet connections. They would like their employees to use the internet only for legitimate business purposes and limit heavy-bandwidth-consumption apps such as video-streaming apps. The **Bandwidth Control** option provides enterprises with such an option.

Bandwidth Classes

Before configuring this control, we need to look at bandwidth classes. A bandwidth class is a collection of a similar set of cloud applications or URL domains. For example, the pre-defined bandwidth class of **Streaming Media** contains the hulu.com and youtube.com domains. Zscaler offers some default pre-defined bandwidth classes and offers the option to create custom bandwidth classes.

To edit pre-defined bandwidth classes or to create a new one, click on **Administration -> Resources > Access Control -> Bandwidth Classes**. A custom bandwidth class can be created by leveraging the existing cloud applications or by entering the URL domains as a list.

On the same **Bandwidth Classes** page, navigate to the **Large Files** tab. Here, enterprise administrators can define the value for the minimum file size for it to be considered a large file. The options are **5 MB, 10 MB, 50 MB, 100 MB, 250 MB, 500 MB**, and **1 GB**.

On the **Web Conferencing Applications** tab, the necessary web-conferencing applications can be selected if they want to be policed using bandwidth control as part of the web-conferencing bandwidth class. The available options are **WebEx, GoToMeeting, Microsoft Live Meeting, CONNECT**, and **InterCall**.

On the last tab, **VOIP Applications, Skype** can be included as part of the VOIP bandwidth class.

Configuring bandwidth controls

Once the bandwidth classes are created and ready, click on **Policy -> Web -> Access Control -> Bandwidth Control** on the Admin Portal. Zscaler offers some pre-defined bandwidth control rules. Enterprise administrators can edit existing, pre-defined **Bandwidth Control** rules or create new ones. When creating a custom **Bandwidth Control** rule, the following options are available:

- **Rule Order**—Similar to the **URL Filtering** or **Cloud App Control** rules earlier, rules are evaluated in the order they are listed. Lower numbered rules are evaluated first, and a first match is selected.

- **Rule Name**—A descriptive name for the rule so that administrators can quickly identify what this rule is for.

- **Rule Status**—The rule status can be enabled or disabled.

- **Bandwidth Classes**—Here, administrators can select one or more pre-defined bandwidth classes or the custom bandwidth classes.

- **Locations**—The location with which this bandwidth control policy needs to be associated. We will look at locations in the *Creating GRE or IPsec Locations* section of *Chapter 4, Understanding Traffic Forwarding and User Authentication Options*. For now, just assume that it could be the New York headquarters location of the enterprise.

- **Location Groups**—Several locations can be grouped for convenience. For example, if an enterprise is in many countries, a location group can be created for each country. For example, the USA Location Group could have the locations of New York, San Francisco, and so on.

- **Time**—The time of the day when this bandwidth control should be applied. Typically, bandwidth controls are applied during business hours so that business-critical application traffic is prioritized over non-critical business application traffic.

- **Protocols**—A collection of one or more types of protocol traffic such as DNS over **HyperText Transfer Protocol Secure** (**HTTPS**), FTP, HTTP, and so on.

- **Minimum Bandwidth**—This value represents the minimum percentage of a location's available bandwidth. For example, if the New York location has a 100 **megabits per second** (**Mbps**) link, a minimum bandwidth percentage of 30% will guarantee at least 10 Mbps traffic for this bandwidth class. This percentage is for both the upload and download bandwidth speeds. This setting kicks in only when there is contention at the location. If business-critical email traffic is getting affected due to bandwidth throttling, bandwidth control kicks in and guarantees a 30% bandwidth for that traffic. This setting is usually used to protect business-critical traffic.

- **Maximum Bandwidth**—This value sets the maximum bandwidth that the bandwidth class can use. For example, if the maximum bandwidth percentage for streaming media is set to 20%, then ZIA will only allow streaming media to consume 20% of the bandwidth in all circumstances, whether there is contention on the link or not. This setting is usually used to limit non-business-critical traffic.

- **Description**—This is a free-flowing text field that can be used to describe the purpose of this bandwidth control rule and it appears on the main **Bandwidth Control** page.

Now that we have explored the bandwidth control options in detail, let's now proceed to discuss **Secure Sockets Layer** (**SSL**) inspection options.

SSL inspection

We talked about SSL inspection in detail in the previous chapter, and the need for it. To configure the SSL inspection parameters, the enterprise administrator can click on **Policy -> Web -> Access Control -> SSL Inspection**. **SSL Inspection** is enabled on a per-location basis, just as with **Bandwidth Control**. Now, let's look at the options available on the **SSL Inspection** page.

If SSL Inspection is Disabled, Block HTTPS to these Sites

This section allows enterprise administrators to specify the policy to apply when **SSL Inspection** is disabled for a location. You can block certain URL categories, using the **Blocked URL Categories** dropdown. You can specify your own list of URLs to block, and the limit is up to 25,000 URLs.

When the **Show Notifications for Blocked Traffic** option is enabled, a message is displayed to the end user that the traffic was blocked due to **SSL Inspection** not being enabled. For the message to be displayed, the Zscaler root certificate must be installed in the end users' browsers. Without the certificate or with this option turned off entirely, end users will only see a message that says **Page Not Found**. This could puzzle end users and make it difficult for helpdesk personnel to troubleshoot.

Policy for SSL Inspection

By default, **SSL Inspection** is enabled for all URL categories, but enterprise administrators have the option to choose only certain URL categories by choosing the **Categories** dropdown under the **Inspect Sessions for These URL Categories** option. Enabling the **Block Undecryptable Traffic** option blocks application traffic that uses non-standard encryption algorithms.

Traffic from certain URL categories can be exempted from **SSL Inspection** by choosing the dropdown under **Exempt These URL Categories from Inspection & Other Policies**. A similar option is available for selected hosts. Those hostnames can be added under the **Exempt These Hosts from Inspection & Other Policies** option. Certain types of cloud applications can also be exempted from **SSL Inspection**, using the dropdown under **Exempt These Applications from Inspection & Other Policies**. Usually, transactions to healthcare and banking URLs are exempted from SSL inspection as those transactions may contain **Personally Identifiable Information** (**PII**) and credit card numbers, bank account numbers, or social security numbers. This depends on the enterprise's employee policy.

There are three options to deal with **Untrusted Server Certificates**. These are certificates from unknown issuers, expired certificates, and certificates where the **Common Name** does not match. The **Allow** option simply allows the traffic, without any errors or warnings. The **Pass-Through** option displays a caution message to end users, and the decision to proceed is left up to the end user. The **Block** option does not allow access to websites with untrusted certificates. Enabling the **Block Site with Revoked Server Certificate** option uses the **Online Certificate Status Protocol** (**OCSP**) to obtain the revocation status of the certificate and blocks the site if that site's certificate is found to be revoked.

Policy for Remote Users with Kerberos

There is only one option under this section: **Enable the SSL Inspection for Remote Users with Kerberos**. Turning on this option performs SSL inspection for roaming users with remote devices using **proxy auto-configuration** (**PAC**) files with Kerberos authentication.

Policy for Zscaler Client Connector

There are four operating system platforms listed in this section—namely, **Windows**, **MacOS**, **Android**, and **iOS**. Enabling these options allows SSL inspection for **Zscaler Client Connector** users on those respective platforms.

Intermediate Root Certificate Authority for SSL Inspection

In the previous chapter, we saw how Zscaler is in between the end user and the destination web server, and the SSL connections terminate on Zscaler. So, the SSL certificate presented by Zscaler to the end user must be trusted by the end user's system. This can be done by downloading the Zscaler intermediate root certificate and populating it to the end users' systems by installing it in the trusted root **certificate authority's (CA's)** store.

An alternate way is to use a custom certificate. The root CA of the enterprise generates an intermediate certificate, and the enterprise administrator then uploads that certificate to the Zscaler Admin Portal here. Since the end users' systems already trust the company's root CA, they will be able to trust these uploaded custom certificates coming from Zscaler that were already signed by the company's root CA.

After exploring the choices available as part of **Access Control**, let's now look at the third major component of ZIA's Web policy, which is **Data Loss Prevention (DLP)**.

DLP

The third and final major component under **Web** is **DLP**. Let's understand what DLP is and why it is so important for today's enterprises. The primary objective of a web or an information security solution is to protect data. That data could be enterprise intellectual property such as a proprietary formula medicine or an algorithm, or valuable data that offers insights on a large segment of customers.

When this valuable enterprise data is stolen by bad actors, it could cause an immense loss to the enterprise in the form of government penalties, lost revenue, lost reputation, and customer loyalty, and, in the worst case, driving the enterprise out of business. So, the loss prevention of this data is critical to the survival of the enterprise.

ZIA DLP solution

Before using a DLP solution, the goals of the enterprise need to be clear. Does the enterprise need active, real-time detection and blocking of data exfiltration, or is it sufficient to adopt a reactive posture and respond after the data loss is detected? Also, the enterprise needs to use clear signatures to mark valuable data so that it can be scanned against traffic leaving the organization.

ZIA offers DLP engines that can be used to scan data against those key signatures for traffic leaving the organization in real time and block it. Just as a common dictionary holds the meanings to words in a language, Zscaler offers 14 standard DLP dictionaries and four default engines. In addition, enterprise administrators can also create up to 32 custom DLP dictionaries based on the business' needs.

Due to the proactive nature of protection, the ZIA DLP solution blocks the data transfer before sensitive data leaves the company. It can then notify the proper company authorities when this transaction triggers a configured DLP rule. All the data related to this occurrence is also logged and sent to the proper storage mechanisms. This data exfiltration is detected not only with HTTP and HTTPS traffic, but also with native FTP traffic.

There is also a reactive mechanism offered by Zscaler. In this case, ZIA only detects (not blocks) the violation and sends it to an on-premises or a cloud-based DLP solution using the **Internet Content Adaptation Protocol** (**ICAP**). Here, the content is not scanned by Zscaler, but only matched based on the configured criteria. Some additional configuration will need to be done by the enterprise to get this data from the Zscaler **Public Service Edge** (**PSE**) in an encrypted manner.

ZIA DLP dictionaries

ZIA DLP dictionaries can be used to detect several varieties of data signatures such as social security numbers, UK National Insurance numbers, and so on. Custom DLP dictionaries can be created using fuzzy search criteria that can match financial or medical data, source code, questionable content, and any other type of specific signature. Data phrases and patterns can be defined in the ZIA dictionaries and can be combined with the Boolean AND operator in the Zscaler engine.

As with the **URL Filtering** and **Cloud App Control** rules, the data context and content can be matched using a policy rule to keywords, patterns, or phrases generated by specific combinations of Users, Groups, Departments, or Locations and within the Time intervals.

When used in conjunction with SSL Inspection, ZIA DLP detects data signatures and applies policies to all traffic regardless of encryption. Web-based email providers such as Yahoo Mail or Gmail, web-based storage solutions such as Box and Google Drive, web-based productivity tools such as Google Apps and Salesforce, and even social media platforms such as Facebook or Twitter can all be inspected.

ZIA DLP configuration

The first step in the configuration process is to review the existing pre-defined DLP dictionaries and engines or create new, custom dictionaries and engines. Enterprise administrators can click on **Administration -> Resources -> Data Loss Prevention -> DLP Dictionaries & Engines**. On this page, there are two tabs, one for **DLP Dictionaries** and one for **DLP Engines**.

Each DLP dictionary also contains a Trigger Threshold. When the number of violations count exceeds the configured Number of Violations Threshold, the dictionary is triggered. The number of violations can be a number up to 10,000. When creating a custom DLP dictionary, you can use a subset of **Portable Operating System Interface (POSIX) Extended Regular Expression (ERE)** to create the patterns that you want to match. You can add up to eight alphanumeric patterns to represent the content you want to protect. You can also add up to 120 exact phrases that you want the dictionary to detect.

A collection of one or more DLP dictionaries makes up a DLP engine. DLP policies make use of DLP engines. There are four pre-defined DLP engine—**HIPAA**, **GLBA**, **PCI**, and **Offensive Language**. If you create a custom DLP engine, you can select a combination of DLP dictionaries that you want the DLP engine to enforce. For a DLP engine to trigger, all its component dictionaries must also trigger.

The second step is to create a DLP notification template. This can be accomplished by going to **Administration -> Resources -> Data Loss Prevention -> DLP Notification Templates**. The options when creating a DLP notification template are as follows:

- **Name**—Use a descriptive name that clearly conveys what this template does.
- **Subject**—This will be the subject of the email that is sent when the notification is sent.
- **Attach Violating Content**—This is an optional setting that, when enabled, attaches the content that triggered the notification.
- **Use TLS**—Uses an encrypted connection when sending out the notification.

You can either choose **Message as Plain Text** or **Message as HTML**. Messages can be customized, and several macros are available for use within the text of the message. Some example macros are ${USER}, ${URL}, ${ENGINES}, and so on. The macros will be substituted with the actual value when the notification is sent.

A third optional step is to set the **Internet Content Adaptation Protocol (ICAP)** settings (**Administration -> ICAP Settings**) if the enterprise plans to send suspicious transactions to an external DLP server using ICAP. **Add ICAP Server** has the following options: the name of the **ICAP Server**, **Enabled** or **Disabled**, and the **Server URI**.

The last step in the configuration process is to define the DLP policies for the enterprise that consist of the engines, templates, and ICAP servers configured in the preceding steps. Navigate to the **Data Loss Prevention** page (**Policy -> Web -> Data Loss Prevention -> Data Loss Prevention**) and click on **Add**. Select **Zscaler DLP Engine**. The options available are listed as follows:

- **Rule Order**—Where in the order this rule resides. Recall that lower-numbered rules are evaluated first.

- **Rule Name**—Enter a name that makes it easy for an administrator to understand what it does.

- **Rule Status**—This can be either enabled or disabled.

- **DLP Engines**—Select a subset of DLP engines already available or created in the previous steps.

- **URL Categories, Cloud Applications**—Select a subset (the default is **Any** for each option).

- **File Type**—Select a subset of the file types (the default is **Any**).

- **Minimum Data Size (KB)**—The minimum size of the data that is matched.

- **Users, Groups, Departments, Locations, Time**—We have already discussed these options.

- **Protocols**—The options are **HTTP**, **HTTPS**, and **Native FTP**.

- **ICAP Server**—You can optionally send this data to an ICAP server.

- **Data Traffic**—Options are to allow traffic or block it. In both cases, the transaction is logged.

- **Auditor Type**—The notification can be sent to a **Hosted** or an **External** auditor.

- **Notification Template**—Choose the notification template that was already created.

You should now have a good understanding of the various components of the ZIA Web policy, the implications of enabling or disabling each option, and the different combinations possible when customizing this policy for your enterprise needs. Let's now explore the options available as part of ZIA Mobile policy.

Exploring the ZIA Mobile policy

Employees of an enterprise are no longer limited to using their desktop or laptop computers. As corporate applications are moving to the cloud, mobile devices such as tablets and smartphones are becoming increasingly popular with the mobile workforce. So, now, along with protecting their traditional computers, enterprises need to secure their mobile infrastructure. We introduced the ZCC in the previous chapter—this can be used by personal computer users by running it as an application on their Windows and macOS computers. The ZCC can also be used as an app on mobile platforms. ZIA offers protection for the iOS and Android mobile platforms.

The ZIA Mobile policy has three major components— **Zscaler Client Connector Configuration**, **Mobile Security**, and **Mobile Access Control**. The ZCC configuration section defines the behavior of the ZCC app for the end users. The mobile security allows an enterprise to set security policies for their end-user mobile devices such as smartphones and tablets. Mobile access control is the equivalent of web access control for mobile devices. Let's now examine these components in detail.

Zscaler Client Connector Configuration

ZCC App can be customized to fit the needs of an enterprise. For example, the end-user ZCC App profiles can be configured to use the **Generic Routing Encapsulation** (**GRE**) or **Internet Protocol Security** (**IPsec**) tunnels when the end user is in a corporate office and automatically switch to a PAC file when they are working remotely from an untrusted location such as an airport or using public Wi-Fi at a coffee shop.

To configure the options for ZCC, enterprise administrators needs to log in to the Admin Portal and click on **Policy -> Mobile -> Zscaler Client Connector Configuration -> Zscaler Client Connector Portal**.

Zscaler Client Connector Portal

The ZCC Portal has a very different look and feel than the main ZIA Admin Portal. A significant difference between the ZIA Admin Portal and the ZCC Portal is that there is no need to activate changes in the ZCC Portal. When you make a change in the ZCC Portal and save the change, it is activated immediately. So, enterprise administrators need to be careful when making any changes on the ZCC Portal that may impact a lot of users.

The main page for the ZCC Portal has four options on the top—**Dashboard**, **Enrolled Devices**, **App Policies**, and **Administration**.

ZCC Dashboard

The Dashboard has some built-in widgets that show the number of licenses in use, the device model, operating systems in use, the device policy status, and the count of devices per platform. The widgets also offer interactive options when the mouse is placed over them. The dashboard can be adjusted using three drop-down menus at the top of the page—**All Users**, **Status**, and **All OS**.

When the **All Users** option is selected, a filter option shows up whereby an administrator can search for a certain user. When that user is selected, the dashboard widgets will update to show the data for that user. This is especially useful when troubleshooting, to see if the user has hit their maximum device limit.

When the **States** option is selected, there is an option to select a certain state. The states available are **Updated**, **Outdated**, **Removal Pending**, **Unregistered**, and **Removed**. Using this filter, an administrator can quickly check if, for example, there are many unregistered devices. By default, **Unregistered** and **Removed** devices are not shown, but can be selected after clicking the **States** option dropdown.

When the **All OS** option is selected, the drop-down options are **iOS**, **Android**, **Windows**, and **MacOS**. Again, an administrator can quickly check details by—say —iOS platform. Let's now explore the **Enrolled Devices** section.

Enrolled Devices

When an administrator navigates to the **Enrolled Devices** section, the page shows **User ID**, **OS Type**, **Device Model**, **ZCC Version**, and **Policy Status** headings in a tabular format. The three filters mentioned earlier (**All Users**, **States**, and **All OS**) can also be used here as well to drill down on the details. A nice feature of this page is that this table can be exported as a **comma-separated values** (**CSV**) file. To quickly check a specific row's detail, click on the page icon to the right of that entry and a popup will appear with details of that registered device.

The popup will have registration details for the device, such as **Device ID**, **Last Registration Time**, **Last Unregistration Time**, and **ZCC Version**. The popup has also a section specifically for the device details, such as **Unique ID**, **Operating System**, **Model**, **Manufacturer**, **MAC Address**, **Device Locale**, and **Hardware Fingerprint**. This information can be used by an administrator to check if a suspicious-looking device shows up on their network.

If an administrator suspects malicious or unauthorized activity, they can remove a registered device from this page. To the right of the device information on the **Enrolled Device Page**, select the checkbox of the suspicious device and click on **Remove Checked Devices**, and click **OK** to confirm. This will deregister that user from that device, clear all saved settings, and revoke all applied certificates. The user will then have to register into the ZCC again to access the network.

Let's now explore the configuration settings available under the **App Profiles** tab.

App Profiles

As soon as the administrator clicks on the **App Profiles** option at the top and navigates to that page, on the left there are two icons, one for **Mobile Devices** and one for **Personal Computers**. The **Mobile Devices** option contains **iOS** and **Android**, and the **Personal Computers** option contains **Windows** and **MacOS**.

An app profile is a collection of configuration settings. Let's look at some of those. The first setting is related to whether end users must enter an administrator-provisioned password to disable or log out of ZCC. Another setting is whether the URL for a custom PAC file is needed for the end user. A third setting decides how the ZCC detects trusted networks and manages traffic forwarding on a trusted or non-trusted network. Some more settings decide if ZCC can install a Zscaler SSL certificate on an end user's devices to allow SSL inspection of traffic handled by ZCC. A last setting determines how ZCC generates logs and the maximum log file size.

As with the **URL Filtering** and **Cloud App Control** rules, several app profiles can be created in the proper order of preference. Based on the precedence, the right app profile is selected for an end user when the user registers with and logs in to ZCC. For example, **App profile 1** is used when a user is working remotely, and **App profile 2** may be used when the same user travels to a corporate location where the user is on a trusted network.

Let's now look at the various options under the **Administration** tab.

Administration

The first option on the left of the **Administration** page is **ZCC App Store**. Under the **Personal Computers** tab, the ZCC application can be downloaded as an executable file (`.exe`) or a Windows Installer file (`.msi`) for the Windows PC platforms and as an executable file for macOS computers. Typically in an enterprise, the end users are not allowed to browse the internet and download the ZCC executable from the open internet. Instead, the enterprise administrator downloads the desired version of ZCC from this portal and works with their IT team to roll it out to the employees' PCs. In such cases, it is recommended to set the **Automatic Rollout** option on this page to **Disable**.

Also, once the ZCC application is installed on the end users' computers, they can click on the **Update App** option to update the version. In that case, the **Automatic Rollout** option can be set to **Always Latest Version** or **Specific Version** and they can choose the desired version on this page. If the **Disable** option is selected, then the enterprise administrator must manually push the newer versions of ZCC through their IT teams from time to time. It is up to the enterprise to make their choices accordingly.

The second option on the left is **ZCC Notifications**. The first tab on this page is **Acceptable Use Policy (AUP) Settings**. Here, an administrator can configure how frequently the AUP is displayed to the end user. The option settings are **Never**, **After each login**, **After device reboot**, **Daily**, **Weekly**, and **Custom**. The default AUP message can also be customized as per the enterprise needs.

The second tab on this page is for **Reminder Notification Settings**. If the ZCC app is inactive, a security reminder can be displayed as per the frequency setting on this tab. The last tab is the **System Tray Notification Settings** tab. There is only one option here, which says **Enable Notifications by Default**. This is enabled by default so that end users can see notifications from the ZCC App.

The third option on the left of the **Administration** page is the **Forwarding Profile** option. Under this section, trusted network conditions and policy configuration can be set for **Trusted Network**, **System Proxy**, **VPN Trusted Network**, and **Off Trusted Network**.

The fourth option on the left is **ZCC Support**. The first tab is titled **App Supportability** and has the following options:

- **Hide Logging Controls on ZCC App**—This option will hide the logging controls for end users.

- **Enable Support Access in ZCC App**—This option allows end users to directly create a trouble ticket from within the app. A helpdesk email address can be configured to send logs when that case is created by the end user.

- **Enable End User Ticket Submission to Zscaler**—When enabled, this allows end users to raise support tickets directly with Zscaler. In a large enterprise, it is recommended to turn off this option.

- **Enable End User to Restart Services and Repair App**—This option enables end users to restart the application or repair it.

The second tab is called **App Fail Open** and it has the following options:

- **If Captive Portal Detected, then Disable Web Security for**—This option is followed by a time setting in minutes. When the end user is in a place that has public internet with a captive portal (such as a coffee shop or an airport), then the ZCC app will wait until the captive portal login is completed before enabling the internet security.

- **If Public Service Edge is not reachable, then**—This is followed by two option settings. The first setting, **Send Traffic Direct**, will send the end-user traffic to the internet, directly bypassing the ZCC App. The second setting, **Disable Internet Access**, will block end users from accessing the internet.

- **If Zscaler Client Connector Tunnel Setup Fails**—This option also has the same two settings mentioned previously.

Another important tab under **ZCC Support** is **User Privacy**. This contains options to enable or disable collection of the device owner and machine hostname information. You can also enable or disable local packet capture in ZCC and enable or disable automatic crash reporting when an app crashes.

The fifth option on the left is **Zscaler Service Entitlement**. This option is enabled only if the enterprise subscribes to the ZPA service as well. The sixth option is **Zscaler App IdP**, whereby silent authentication into the ZCC can be enabled that uses the user's device login. The last option is **Device Posture**, which again is applicable only to the ZPA service.

After examining the options for **Zscaler Client Connector Configuration**, let's now move on to the options available as part of **Zscaler Mobile Security**.

Zscaler Mobile Security

The second major component of the Zscaler Mobile policy is Zscaler Security for Mobile devices. Administrators need to click on **Policy -> Mobile -> Security -> Mobile Malware Protection**. Let's explore the options available here.

Mobile App Security Actions

Here, the administrator can **Allow** or **Block Malicious Activity** and **Known Vulnerabilities**.

Mobile App Privacy Options

All the following options for mobile apps can be set to **Allow** or **Block**. Mobile apps that send users' credentials without encryption, end users' location information, end users' PII, device identifiers, communicate with Ad servers, and communicate with unknown servers.

The last component of the Zscaler Mobile policy is the **Zscaler Mobile Access Control** component. Let's explore the options available for this.

Zscaler Mobile Access Control

The third and last component of the Zscaler Mobile policy is **Zscaler Mobile Access Control**. Administrators can navigate here by clicking on **Policy -> Mobile -> Access Control -> Mobile App Store Control**. Rules can be created in the order of preference under this section, just as with the **URL Filtering** and **Cloud App Control** rules. Each rule contains the following information:

- **Rule Order**—Where this rule is placed with respect to the other rules.
- **Rule Name**—A descriptive rule name.
- **Rule Status**—This can be either enabled or disabled.
- **App Stores**—The default selection is **Any**, but an administrator can choose certain stores such as Google Play or Android Marketplace.
- **Users, Groups, Departments, Locations, Location Groups, Time**—These options have already been explained before.
- **Action**—If this action is explicitly allowed or blocked.
- **Description**—A free-flowing text field that contains a detailed description of the rule for quick reference on the main **Mobile App Store Control** page.

After exploring the ZIA Mobile policy options in detail, it is now time to look at the third and last major component of the ZIA policy, which is the ZIA Firewall policy.

Learning about the ZIA Firewall policy

Traditionally, physical firewalls have been in use for a very long time by enterprises. Firewalls limit unsolicited inbound traffic to an enterprise and keep track of the outbound data connections generated by end users and applications, and only allow them back in. These firewalls also perform additional functions such as NAT and protocol inspection.

Zscaler offers a basic and advanced cloud firewall capability that allows configuration of access control policies, as with a physical firewall. The basic firewall only allows you to create rules using source IP address, destination IP address, source port, destination port, and protocol. The advanced firewall offers the use of Zscaler default Network Services and Network Application definitions. It also allows the creation of custom services.

The Firewall policy consists of five main components—**Firewall Control**, **NAT Control**, **DNS Control**, **FTP Control**, and **IPS Control**. Let's now start by exploring the **Firewall Control** component.

Firewall Control

After logging in to the Admin portal, administrators can go to this section by clicking on **Policy -> Firewall Filtering > Access Control -> Firewall Control**. On this page, there are two tabs, **Firewall Filtering Policy** and **NAT Control Policy**. Let's look at the first tab now.

Firewall Filtering Policy

By default, Zscaler has two built-in firewall filtering rules. The first one allows Office 365 traffic for all users and the default firewall filtering rule that allows any traffic. The default rule can be edited to block all the traffic by dropping the traffic silently, or by sending an **Internet Control Message Protocol** (**ICMP**) destination unreachable message to the client, or a **Transmission Control Protocol** (**TCP**) connection reset (**Block** with **Drop** for non-TCP traffic). Once the enterprise has configured and tested its desired firewall rules, then this default filtering rule can be edited to drop the traffic as desired.

To add a new Firewall filtering rule, click on the blue + icon. On the pop-up menu, there are many options, as follows:

- **Rule Order**—Just as with a regular firewall, make sure the most specific rules are at the top of the list of rules and that the least specific are at the bottom of the list.

- **Rule Name**—A descriptive name for the rule that is easy to understand.

- **Rule Status**—This state could be enabled or disabled.

- **Who, Where, & When**—The already-familiar options are **Users**, **Groups**, **Departments**, **Locations**, **Location Groups**, and **Time**.

- **Network Traffic**—The action to take for this traffic. The default option setting is **Allow**, but it can also be set to **Block** with the **Drop**, **ICMP**, and **Reset** options already explained previously.

- **Description**—A free-flowing text field to describe the purpose of this rule in a business context.

In addition to the **Who, Where, & When** options, there are three more tabs, as follows:

- **Services & Applications**—The criteria here are **Network Service Groups**, **Network Services**, **Network Application Groups**, and **Network Applications**. The default option setting for the groups is **None**, and **Any** for the others.

- **Source IP**—Source IP groups that were already defined can be selected, or IP addresses can be added to the list.

- **Destination IP**—Destination groups that were already defined can be selected and IP addresses, or **Fully Qualified Domain Names** (**FQDNs**) can be added. You can also select the destination countries and URL categories.

As you can see, this process is very similar to filtering rules that are added to a typical firewall device. Now, let's examine the **NAT Control Policy** section.

NAT Control Policy

NAT is used to translate one IP address range and/or port to another IP address range and/or port. This is commonly used in our homes, where our ISP router translates the public-facing IP address to the private IP address range on the internal home network so that multiple devices can connect to the internet.

Administrators can go to this section by clicking on **Policy -> Firewall Filtering -> Access Control -> NAT Control**. There are no default NAT control rules. Click on the blue + icon to add a new NAT control rule. The options are very similar to the **Firewall Filtering** rules. There is no **Action** section for NAT control rules. The other few differences are explained here:

- **Services**—Instead of **Service & Applications**, you just have the **Services** option here. It includes only **Network Service Groups** and **Network Services**.

- **DNAT IP Address or FQDN**—Traffic that matches the conditions under the **Criteria** section will be sent to this destination.

- **DNAT Port**—Traffic that matches the conditions under the **Criteria** section will be sent to this port.

Let's now examine the **DNS Control** component of the Zscaler Firewall policy.

DNS Control

Enterprise administrators can reach this section by clicking on **Policy -> Firewall Filtering -> Access Control -> DNS Control**. **DNS Control** has three default **DNS Filtering Allow** rules. The first one is for Office 365 traffic. The second one allows all unknown DNS traffic, and the third and last one allows any traffic. As with the default firewall filtering rule, once the enterprise has its preferred **DNS Filtering** rules in place, they should modify the second and third default **DNS Filtering** rules.

A new **DNS Filtering** rule can be added by clicking on the blue + icon. The options are very similar to the **Firewall Filtering** rules, with minor differences explained here:

- **Destination/Resolved IP**—The criteria here are **DNS Server IP Groups** (already configured previously) or a list of **DNS Server IP Addresses**.

- **DNS Application**—The criteria here are **DNS Tunnels & Network Apps, DNS Application Group, Resolved IP-Based Controls, Requested Domain/Resolved IP Categories**, and **DNS Request Type**. The default option settings for all of these is **Any**.

- **Action**—The action has different options for handling network traffic. The default is **Allow**. The second option is **Block**. The third option setting is **Redirect Request**. When selected, it prompts for the **DNS Server IP Address**. The fourth and last option is **Redirect Response**, which prompts for the **IP Address**.

Now, let's look at third component of the Zscaler Firewall policy, which is the **FTP Control** component.

FTP Control

Administrators can navigate to this section by clicking on **Policy -> Firewall Filtering -> Access Control -> FTP Control**. Administrators can decide whether to enable or disable the **Allow FTP over HTTP traffic** and the **Native FTP traffic** settings.

Let's look at the **IPS Control** component now.

IPS Control

Administrators can visit this section by clicking on **Policy -> Firewall Filtering -> Access Control -> IPS Control**. The default IPS rule is set to **Block all traffic by silently dropping it**. **IPS Control** has very similar options to the **Firewall**, **NAT**, and **DNS** rules. The differences are explained here:

- **Services & Threats**—The criteria in this section are **Network Service Groups**, **Network Services**, and **Threat Categories**.

- **Action**—The default action is **Allow**. The other options are **Block by silently dropping traffic**, **Block and Reset the TCP connection**, and **Bypass IPS**.

You should now have a strong understanding of the various options offered by the Zscaler Firewall policy. It is also important to visualize the order in which these policies are enforced by the Zscaler platform. Let's review that now.

Order of policy enforcement

When an end user generates a web request and when that request reaches the Zscaler PSE, the PSE first enforces the **Browser Control** policy. If the end user's browser does not meet the enterprise requirements, that request will be blocked. The **FTP Control** policy acts next, based on the policy configuration. It is followed by the **URL and Cloud App Control** policy, and then the **SSL Inspection** policy.

If the request satisfies the conditions mentioned previously, it is next enforced by the **Malware Protection** policy, followed by the **Advanced Threat Protection** policy, the **Upload File Type Control** policy, the **Upload Bandwidth Control** policy, the **Data Loss Prevention** policy, and finally, the **Firewall and DNS Control** policy. If the request survives all these policies, it can proceed.

When the response to that initial end user request arrives at the Zscaler PSE, the policy order is reversed. The response goes through the **Inbound Malware Protection** policy, the **Inbound Advanced Threat Protection** policy, the **Sandbox** policy, the **Download File Type Control** policy, and, finally, the **Download Bandwidth Control** policy. If the response satisfies all these requests, it can proceed to the end user who originated the request.

Summary

In this chapter, we looked at the features of the ZIA policy that can be broadly categorized into Web, Mobile, and Firewall policies. An enterprise can customize these various policies to suit its business needs. A strong understanding of the implications of turning on a certain option is necessary for an enterprise administrator to create a robust security policy necessary for an enterprise to meet its security obligations. This knowledge will come in handy when architecting a Zscaler ZIA solution.

In the next chapter, we will look at the various options of forwarding end-user traffic to Zscaler, the authentication methods that can be configured for end users, and the capability and flexibility of the **Zscaler Client Connector** application.

Questions

As we conclude, here is a list of questions for you to test your knowledge regarding this chapter's material. You will find the answers in the *Assessments* section of the Appendix:

1. A ZIA policy is broadly categorized into the following sections:

   a. Firewall

   b. Web

   c. Security

   d. All the above

2. ZIA does not allow for SSL Inspection exemption.

   a. True

   b. False

3. The **Zscaler Client Connector** application can be used on the following platforms:

   a. Windows and macOS computers

   b. Android and iOS devices

   c. All the above

   d. None of the above

4. The default ZIA Firewall filtering rule blocks all traffic.

    a. True

    b. False

Further reading

- Office 365 options: `https://help.zscaler.com/zia/about-microsoft-one-click-options`

4
Understanding Traffic Forwarding and User Authentication Options

In addition to the **Zscaler Internet Access (ZIA)** policy, understanding traffic forwarding and user authentication options answer the important questions that allow an enterprise to plan their migration to a ZIA solution. As part of these two topics, we will explore the **Zscaler Client Connector (ZCC)** in more detail. Recall that ZCC is an application that offers tremendous flexibility to enterprise end users.

We have already looked at the various ZIA features that can be applied to user web traffic in depth. Next, we will need to understand how the users' web traffic gets to Zscaler and how they can authenticate. We already explained that user authentication enables us to identify the end users and apply granular policies at the user, group, department, and location level.

After learning about these options, we will explore the approach of discovering the current security posture of the enterprise by asking the right questions. The answers that are derived from the stakeholders can then give shape to what a ZIA solution might look like.

In this chapter, we are going to cover the following main topics:

- Understanding traffic forwarding
- Exploring ZCC internet traffic forwarding
- Evaluating user authentication options

Let's get started!

Technical requirements

A familiarity with technologies such as GRE and IPsec tunnels, as well as how PAC files work, will be required for this chapter.

Understanding traffic forwarding

There are different ways to send end user web traffic to Zscaler. The recommended and well-supported options are using GRE or IPsec tunnels, PAC files, and the ZCC application. The proxy chaining option is not recommended for production or long-term use. Other limited or deprecated choices include using port forwarding and secure agents.

The recommended choices for corporate locations are GRE, IPsec, PAC files, or the ZCC application. For mobile users, only PAC files and the ZCC app options are available. We will only discuss the recommended options here. First, we'll look at the GRE tunnel forwarding option.

GRE tunnel forwarding

Generic Routing Encapsulation (GRE) is a tunneling protocol that can be used to encapsulate data packets using one protocol inside the packets of another protocol. GRE enables us to use protocols not usually supported by a certain network because now, those packets are put inside other packets that use supported protocols.

GRE tunnel traffic forwarding is the most common method That is used by many enterprise office locations. In this case, GRE tunnels are created from the edge device (such as a router or a firewall) located on the enterprise's premises to the nearest or chosen Zscaler Public Service Edge. Typically, for high availability, one tunnel is provisioned to the primary Zscaler PSE and one tunnel is configured to the secondary Zscaler PSE. Let's consider the requirements for GRE tunnels.

The first key requirement is that the edge device should have a routable and static public IP address. The enterprise must make sure from their service provider, such as an ISP, that the IP address provided by them meets this requirement. If the GRE-capable device is behind a firewall performing NAT, then two routable and static IP addresses are needed for each location.

The second key requirement is to have a compatible device. The enterprise needs to check the vendor of the device and make sure it qualifies.

Once both requirements are met and the edge device is fully in place, the enterprise administrator should ideally choose at least two Zscaler PSEs (for redundancy) that yield the best latency and performance values from the edge device. They can then contact Zscaler support with the public IP address and the two PSE choices and put in a provisioning request. If the preferred PSEs are not provided by the enterprise, Zscaler provisioning support will provision the best PSEs according to their internal tools. Note that the enterprise can provision at least one tunnel and add more per their business redundancy requirements. Two tunnels is just a minimum, keeping in mind the need for redundancy.

Once Zscaler support provisions this public IP, the following IP addresses are provided to the enterprise administrator. This information is available both in the text of the Zscaler provisioning ticket and on the Admin Portal:

IP Address	Description
Tunnel Source IP	This is the public IP address of the enterprise-owned GRE device.
Primary Destination	This is the IP address of the primary Zscaler PSE.
Internal Router IP	This is the source (near-end) IP address configured on the primary GRE tunnel on the enterprise-owned GRE device.
Internal ZEN (or PSE) IP	This is the destination (far-end) IP address configured on the Zscaler PSE for the primary tunnel.
Secondary Destination	This is the IP address of the secondary Zscaler PSE.
Internal Router IP	This is the source (near-end) IP address configured on the secondary GRE tunnel on the enterprise-owned GRE device.
Internal ZEN (or PSE) IP	This is the destination (far-end) IP address configured on the Zscaler PSE for the secondary tunnel.

The enterprise administrator then proceeds to configure the primary and secondary GRE tunnels on their GRE edge device. Optionally, if the enterprise wants to use automatic failover, it can configure IP SLA based on packet loss and latency. Once this has been configured, the end user traffic automatically switches to the secondary tunnel whenever the primary tunnel experiences trouble. When the issue on the primary tunnel is resolved, traffic will switch back from the secondary tunnel to the primary tunnel automatically.

If the enterprise wants to add an additional layer of high availability, it can provision two or more GRE-capable devices in an active-standby configuration, with two or more GRE tunnels on each device. If two devices are provisioned with two tunnels on each device, the enterprise will then have four GRE tunnels to the Zscaler cloud. In an extreme business case, this setup can be replicated in a secondary data center, giving the enterprise four GRE-capable devices and eight GRE tunnels!

IPsec tunnel forwarding

Internet Protocol Security (IPsec) is a suite of secure network protocols. It can authenticate and encrypt the data packets between two endpoints, such as computers providing encrypted communications. It is primarily used in **virtual private networks (VPNs)** and provides confidentiality, integrity, and authentication – commonly known as the CIA triad.

IPsec tunnels are another commonly deployed solution by enterprises for their corporate locations. There are two modes supported – aggressive mode and main mode. Aggressive mode uses dynamic IP addressing and a VPN edge device that is compatible. Main mode uses a routable and static IP address, and the enterprise administrator needs to request the provisioning capability through a Zscaler support ticket, very similar to the GRE tunnel scenario.

Like the redundancy seen in the case of GRE tunnels, at least two IPsec VPN tunnels should be configured for each enterprise corporate location. Now that we have seen the GRE and IPsec options, let's see how a location is created for either of these options.

Creating GRE or IPsec locations

Once the GRE or IPsec tunnels have been provisioned by Zscaler support, recall that the enterprise administrator configures this information on their edge device. In addition to this configuration, the administrator also needs to log into the Admin Portal and configure it there for the tunnels to come up and start working on the edge device.

To create a new location or manage existing locations, navigate to **Administration -> Resources -> Traffic Forwarding -> Location Management**. Click on the **Add Location** link. Let's explore the various subsections you'll come across when creating a new location.

Location

The following fields are included for a location:

- **Name**: This text field can be used to indicate the location of the tunnel (where the enterprise edge device is located). It is extremely important that the naming convention is designed properly, especially if the enterprise has a lot of corporate locations. If there is more than one GRE-capable device at the same location, include the device name here.

- **Country**: This is the country where the edge device is located.

- **City/State/Province**: Each country has its own convention when it comes to its city, state, province, or county. This text field can be leveraged accordingly.

- **Time Zone**: The time zone of the location where the edge device is located.

- **Manual Location Groups**: Zscaler allows you to create location groups. Each location group can be a collection of similar locations, such as all the locations in United States, France, and so on. Alternatively, it can be a collection of locations in a state such as California, North Carolina, and so on. Location groups can be useful when you're defining the scope of a new or existing enterprise administrator and can also be used within policies and reporting. For example, instead of adding several locations within a country, just the location group for that country can be used.

- **Dynamic Location Groups**: This is very similar to the case described previously. Instead of manually creating a location group, dynamic location groups can be created using conditions such as City, Country, or Name. This location will show up as a member of a dynamic location group if it matches the conditions that have been defined for a dynamic location group.

- **Exclude from Manual Location Groups**: When enabled, this option prevents this location from being added to a manual location group and will be removed from a manual location group if it's already been added.

- **Exclude from Dynamic Location Groups**: Provides a similar behavior to what was described in the previous bullet point.

Let's look at the next section, which determines whether this is a GRE or an IPsec location.

Addressing

The two choices under this section are as follows:

- **Static IP Addresses**: This is the public IP address of the edge device that was provided to the Zscaler provisioning team by the enterprise administrator. In addition to the configuration on the edge device, recall that the enterprise administrator needs to select that IP address here when creating a location.

- **VPN Credentials**: Choose the appropriate FQDN that identifies the correct VPN peer. This is applicable when you're using IPsec tunnels.

Once it has been decided whether this location will use either GRE or IPsec tunnels, we need to look at the options for this gateway.

Gateway options

The choices available for this section are as follows:

- **Use XFF from Client Request**: If the enterprise is using proxy chaining to forward end user traffic to Zscaler, this option can be enabled so that Zscaler can use the XFF headers that the on-premises proxy server inserts into the outbound HTTP requests.

- **Enforce Authentication**: When enabled, this option will require the end user to authenticate using the authentication mechanism defined by the administrator. We will look at the authentication options that are available in detail elsewhere in this chapter.

- **Enable Caution**: If the previous option is disabled, this option can be enabled, and the caution interval can be set to more than 60 seconds to display a caution message to the unauthenticated users.

- **Enable AUP**: If enabled, this option shows an acceptable user policy to unauthenticated end users and requires them to accept it before proceeding further.

- **Enable SSL Inspection**: We talked about SSL inspection in detail in the previous chapters and the need for it. This option, when enabled, turns on SSL inspection for all the traffic from this location.

- **Enforce Zscaler Client Connector SSL Setting**: This setting, when enabled, uses the SSL Inspection setting for ZCC traffic and overrides the location's SSL Inspection policy for all ZCC traffic.

- **Enforce Firewall Control**: When enabled, this option turns on the firewall for this location.

The last option is the option to either enable or disable bandwidth control.

Bandwidth control

There are only two choices for the **Enforce Bandwidth Control** option. When enabled, this turns on the bandwidth control policies for the traffic from this location.

> **Important Note**
> Once locations have been created, sub-locations can be created under those locations that use IP addresses encapsulated in a GRE or IPsec tunnel or that are sent to Zscaler using XFF headers. Sub-locations are useful where different policies can be implemented based on IP addresses.

Now that we have looked at all the available choices when it comes to creating a new location, let's look at the PAC file forwarding options.

PAC file forwarding

A **Proxy Auto Configuration (PAC)** file is a file that uses JavaScript notation and tells the end user browser which proxy to connect to, over which port, and which destinations or protocols that need to be bypassed and the traffic to those bypassed destinations to be sent directly.

It is recommended that the enterprise hosts this PAC file on a PAC server in the Zscaler cloud. There are some advanced options, such as geo-IP proxy configuration, that are available in this scenario. This does require the end user device to be allowed, by the enterprise firewall, to connect to the Zscaler PAC server and download the PAC file. If the enterprise has business or compliance requirements, they can always host the PAC files internally.

When the PAC files are hosted in the Zscaler cloud, the geo-IP proxy configuration option mentioned previously can be used with the help of Zscaler variables, namely ${GATEWAY} and ${SECONDARY_GATEWAY}. When an end user's browser downloads this PAC file, Zscaler automatically inserts the IP addresses of the closest and most optimal PSEs. When the PAC file is eventually interpreted by the browser, end user traffic is automatically sent to the best PSE. If both the PSEs are somehow unreachable, a catch-all third option called DIRECT tells the browser to send the end user traffic directly to the internet.

In this section, we learned about the different traffic forwarding options available with ZIA. Depending on the enterprise needs, the administrator can choose one or more options. The recommended traffic forwarding option by Zscaler is a combination of GRE tunnels and PAC files.

Now, let's look at the various traffic forwarding options offered by the ZCC.

Exploring ZCC internet traffic forwarding

ZCC is the flagship product of Zscaler. It is central to secure internet access, especially for mobile users, and for secure private application access using **Zscaler Private Access (ZPA)**, which we will explore in the later chapters. Due to the versatility of ZCC, we decided to address it under a section of its own!

Let's look at the traffic forwarding scenarios, as well as the options and authentication options available with ZCC.

ZCC internet access forwarding scenarios

The forwarding profile tells the ZCC app how to treat the ZIA and ZPA traffic coming from the end user systems in various network environments. There are three such forwarding scenarios when it comes to using the ZCC app for secure internet access. Let's look at them in detail.

Remote users

As we discussed in this chapter's introduction, in this case, the remote user is connecting from an untrusted location such as a coffee shop, an airport, or the user's home. Based on our discussion in *Chapter 1, Security for the Modern Enterprise with Zscaler*, many enterprise users fall into this category. In this case, ZCC tunnels or proxies the traffic from the end user device to the Zscaler cloud, thus providing secure internet access.

Trusted network

When these remote users travel to a trusted location such as their corporate office, they are said to be on a trusted network. Typically, these enterprise locations will establish a GRE or an IPsec tunnel to the Zscaler Public Service Edges. At a trusted location, all the network components are typically protected, and the traffic will be scanned. In this case, ZCC can detect the trusted network and use the appropriate forwarding options.

VPN trusted network

In this scenario, a remote user establishes a VPN connection to their enterprise's trusted network. In a full-tunnel scenario, the VPN carries all traffic, in which case the ZCC enforces appropriate traffic forwarding options. In a split-tunnel scenario, the VPN carries non-web traffic such as traffic to the datacenter for legacy applications, and the remaining web traffic goes directly through the end user's local internet to the nearest PSE via the ZCC application.

Now that we have reviewed the three traffic forwarding scenarios for internet access via ZCC app, let's look at the internet forwarding options offered by the ZCC app.

ZCC internet access forwarding options

There are three main forwarding options for internet access. Let's try to understand these options.

Tunnel 1.0

This forwarding option uses lightweight, unencrypted HTTP tunnels on destination port 443 to the nearest or configured Zscaler Public Service Edge on the ZIA cloud. This option is used to forward all end user traffic on ports 80 and 443. This is not limited to browser-generated traffic, but all application traffic that uses ports 80 and 443.

The forwarding profile allows you to use a system proxy that's been locally defined in the browser settings. The option settings are **Enforce**, **Never**, or **Apply on a Network Change**. The proxy settings may be detected automatically, applied using a script, set to a local proxy server on the LAN, applied using a GPO update, or a combination of any of these options.

The App Profile allows you to specify an IP address or a hostname bypass for the VPN gateway. IP addresses or FQDNs used by the enterprise's corporate VPN gateways can be specified here, which effectively tells the ZCC not to process the traffic to these destinations. You also have the choice to apply a custom PAC file, which may be used for two reasons. The first reason is to specify a Service Edge (either a Public or a Private one) for any compliance or preference purposes. The second reason is to specify destinations, the traffic to which is not sent to the Zscaler cloud at all.

Tunnel 1.0 with Local Proxy

This Tunnel with Local Proxy method also sets up a lightweight HTTP CONNECT tunnel but uses a loopback IP-based socket. In this mode, all HTTP and HTTPS traffic, regardless of the port, is sent to Zscaler. All other traffic is sent directly to the internet, bypassing the ZCC. This option can only be used with the Tunnel 1.0 forwarding mode.

Internally, this HTTP and HTTPS traffic is forwarded to the loopback address on port 9000 by default and this port number can be changed on the ZCC Admin Portal.

Tunnel 2.0

This mode forwards all IPv4 unicast traffic on any port. The ZCC creates a TLS tunnel to the nearest healthy or configured Public Service Edge that is being used as a control channel. DTLS, TLS, or Tunnel 1.0 data tunnels are then set up, depending on the configured settings. The preferred order is DTLS, TLS, and then Tunnel 1.0. The preferred tunneling behavior and fallback process can be set in the Forwarding Profile. You can also specify the **Maximum Transmission Unit (MTU)** if the administrator encounters any fragmentation issues.

> **Important Note**
>
> ZCC version 2.0.1 or higher is needed for the Tunnel 2.0 forwarding option to work. On PC, a Packet Filter Based driver is needed on Windows or a ZCC App installed TUN driver on MacOS.

Just like in the previous forwarding mode, the VPN gateway bypass and PAC file settings can be configured in the App Profile. There are also additional options you can use to include or exclude specific networks. These are applied in the following order: VPN gateway bypasses, Tunnel 2.0 Inclusions/Exclusions, and finally, the PAC file bypasses. An understanding of their exclusions and inclusions is critical, especially when it comes to migrating from Tunnel 1.0 to Tunnel 2.0.

Enforce proxy

As the name of this mode suggests, no tunnels are created or used by the ZCC. Instead, the ZCC app simply enforces either the PAC file configured on the end user's browser or a PAC file specified by the IT administrator. If the end user tries to change the proxy settings, they will be reversed by the ZCC app and the specified PAC file is reapplied. The applied PAC file can be used to define the traffic destined for the Zscaler cloud and the traffic that needs to be bypassed. In summary, the ZCC app plays no part in actual traffic forwarding, but simply enforces the system-specified PAC file.

None

When you use the **None** option, the ZCC app does not forward any end user traffic to the Zscaler cloud. There is an option, however, to never apply a proxy configuration or apply one to a network change. If the network changes, the same proxy settings discussed earlier with the Tunnel methods are available.

So far, we have investigated the ZCC traffic forwarding modes and options. Now, let's explore the silent authentication option offered by ZCC.

ZCC silent authentication

To offer an effortless end user experience, the enterprise administrator can enable silent user authentication with the ZCC app. This process involves the following steps:

1. The administrator logs into the Zscaler Admin portal and then navigates to **ZCC App Portal** and creates device tokens (maximum of 8). These can be used when you're installing the app.

2. The administrator then enables the Add Zscaler App Portal under the **IdP** option in the ZCC Admin Portal, under **Administration -> Authentication Settings -> Identity Providers**, optionally choosing **SAML Auto-Provisioning**.

3. The ZCC app is then installed with both the --deviceToken and --userDomain command-line options.

4. After installation, the ZCC app reads the end user's identity from the end user device's login.

5. Using that user identity (*step 4* here) and the provisioned Device Token (*step 3* here), the ZCC app authenticates the end user silently to the Zscaler IdP. The end user neither needs to respond to the ZCC app nor enter a password to enroll.

> **Important Note**
> This silent authentication option is not supported on all mobile platforms and can only be used on managed iOS devices. This method is only applicable to ZIA, not ZPA.

Now that we've reviewed the steps involved in setting up silent authentication in the ZCC app, let's look at a high-level process flow for internet access while using the ZCC app.

ZCC ZIA process flow

The high-level process flow for secure internet access using the ZCC app involves the following steps:

1. The enterprise administrator configures the enterprise-chosen settings for users and/or groups in App Profiles and, optionally, one or more Forwarding Profiles. Additional configuration choices are settings such as ZCC App Notifications, Trusted Networks, ZCC App Support, User Agent, and Zscaler App IdP configurations (explained in the previous section).

2. The administrator works with the IT team that distributes and installs the app for the necessary end users.

3. The end user enrolls into the ZCC app. This can be done using the silent option, which we explained in the previous section.

4. Once the user enrollment has been successful, the app is provisioned and configured by the matching profiles. This includes provisioning the digest credentials necessary to establish the Z tunnels.

5. The ZCC app sends the end user device information, which includes the device's fingerprint for security and reporting. The app also checks every 15 minutes for App Profile and Forwarding Profile PAC file updates; every hour for profile and/or policy updates and to refresh the device fingerprint; and every 2 hours for new software versions. The user can also manually force a check-in for policy or PAC file updates from within the app.

6. End user traffic is then sent by tunnel or proxy, depending on the forwarding mode set by the applied Forwarding Profile.

7. Traffic for the bypass destinations defined and deployed to the app using a custom PAC file will be sent directly without the ZCC App processing it. The app checks every 15 minutes for App Profile PAC file updates (as mentioned in *Step 5*) and caches that file so that the local bypass sites can still be reached in the case of an internet outage.

So far in this section, we have looked at the various traffic forwarding methods offered by ZIA, including the usage of the ZCC app. Let's explore the various authentication options that are available for end users in ZIA.

Evaluating user authentication options

We saw how end user internet traffic is forwarded to Zscaler in the previous sections. When creating a new location, we saw an option called **Enforce Authentication**, which can force the end user to authenticate. If an end user is not authenticated, the administrator has no option but to apply all the policies on a per-location basis. If the policies are applied on a per location basis, there is no visibility into user-level web activities. This choice does not fully leverage the granular policy application feature of ZIA.

Before we evaluate the available user authentication options, let's examine why enterprises usually require user authentication.

Benefits of user authentication

It is very common for end users to authenticate into any application to gain access. Consequently, their permissions are granted accordingly. Similarly, when users authenticate into ZIA, policies can be set on a granular level based on the user, user's group, or department. For example, a user in the social media outreach department needs upload (POST) and download (GET) access to social media platforms, but users from other departments are only granted download (GET) access to the same social media platforms.

Also, consider this situation. If we were applying only location-based policies, what if an enterprise end user travels from their usual United States location to a European Union location? The user experience will differ in the new corporate location. If a user-level policy was used instead, then it will take precedence over the location-based policy, and the user experience remains the same.

Surrogate IP

ZIA offers an option called surrogate IP that can map the end user's device IP address to the user itself. It can be used instead of location-based policies for traffic where user authentication is not possible. This option can be used to authenticate unknown user agents and even non-HTTP protocols.

User provisioning

Let's explore the steps involved in setting up user authentication for Zscaler. The first step is to provision the user and group membership information in the Zscaler Central Authority. This provisioning process depends on the method of authentication that's used by the enterprise.

The second step is where the end user performs the actual authentication so that the Zscaler service can identify the user's identity and apply the policies at a granular level.

User authentication process flow

When the end user traffic reaches the Zscaler cloud, the Zscaler service checks if the traffic is from a known location. If it is, it checks if authentication has been enabled for that known location. If authentication has not been enabled, it applies the location-based policies for that traffic.

If user authentication is enabled for the known location, the end user is prompted to authenticate. If user authentication is successful, user-based policies are applied. If user authentication fails, location-based policies are applied as a fallback mechanism.

If the end user traffic is received from an unknown location, the end user must authenticate into Zscaler to be granted access, and then user-based policies are applied. If the end user fails to authenticate from this unknown location, access is denied.

User authentication options

Zscaler supports four types of user authentication: **Hosted Database (DB)**, **LDAP**, **SAML**, and **Kerberos**. Let's explore each of these options in detail.

Hosted DB

This is the simplest authentication method and is completely supported and managed by Zscaler. Under this option, the enterprise administrator creates individual user accounts using the Admin Portal. Alternatively, a list of users can be entered into a **comma-separated (CSV)** file and uploaded using the same Admin Portal.

This method of authentication uses cookies, and the user authentication frequency can be set to daily, only once, once per session, or a custom time interval. This method is suitable for small enterprises that have a limited set of users.

Let's look at the steps involved in actually configuring the authentication settings.

Configuration steps

After logging into the Admin portal, the enterprise administrator needs to click on **Administration -> Authentication -> Authentication Configuration -> Authentication Settings**. Under the **Authentication Profile** tab and section, select **Hosted DB**. Under the **Authentication Frequency** drop-down menu, select the desired option. If the **Custom** option is chosen, another dropdown is offered that says **Custom Authentication Frequency (days)**. The range can be from 1 to 180 days.

Under **Authentication Type**, choose **Form-Based**. If **Temporary Authentication** is desired (set to **Disabled** by default), choose either the **One-Time Token** or **One-Time Link** option. **Password Strength** can be set to the desired option. depending on the enterprise security policy. The password expiry settings are 1, 3, or 6 months or Never.

Adding individual users

Once the authentication type has been set to **Hosted DB**, the administrator can add individual users. On the Admin Portal, navigate to **Administration -> Authentication -> Authentication Configuration -> User Management**. Before proceeding to create users, go to the **Groups** tab to create or edit groups and the **Departments** tab to create or edit departments. Once you've done that, click on the **Add User** option on the **Users** tab. This will display the following options:

- **User ID**: Enter a unique user ID for the end user. The suffix will be @domain.com for this user. The username for the user will be userID@domain.com.

- **User Display Name**: Enter the user's full name (typically, their first name, middle name, and last name).

- **Groups**: Perform group(s) assignment for this user. A user can be added to up to 128 groups.

- **Department**: Add the user to the appropriate department (only one).

- **Comments**: This is a free-flowing text field that can contain important information about this user.

- **Password/Confirm Password**: It is recommended to use a password generator application to create complex passwords and use those here.

- **Temporary Authentication Email**: This option only appears when the **One-Time Token** or **One-Time Link** option was set earlier. A valid email address can then be entered in this field for temporary authentication.

Adding individual users one by one this way can be a tedious and repetitive job. There is also an option to add or delete multiple users at once. Let's look at that option.

Adding multiple users at once

Adding individual users may become too tedious for the administrator. In such a case, the CSV option mentioned earlier can be used. On the **Users** tab, click on the **Sample Import CSV** file. Open the CSV file. The + symbol in the **Action** column will add the user to that row with the provided parameters. Similarly, the minus symbol will delete the user. Only the Email-ID is necessary for the deletion action.

Once the CSV file is ready, click on the **Import** link on the **Users** tab. In the pop-up menu, if the **Override Existing Entries** checkbox is selected, an existing user enter is overwritten with the new information for the user. If this option is not selected and an already existing user is added, an error message will prevent the import. Once all these settings are ready, click on **Choose file**, import the CSV file, and validate that the desired result has been achieved.

User authentication flow

The end user requests a web page and Zscaler receives the request and asks the end user for their username. When the user enters their username, Zscaler asks them for their password. When the end user enters their password, Zscaler Central Authority compares it with the hosted database and, upon successful authentication, notifies the Zscaler PSE that the end user has been authenticated. It then sends an authentication cookie to the end user browser. Finally, the user is redirected to the web page that was initially requested. This entire authentication process occurs over SSL, and data is always encrypted and never sent in cleartext.

Now that we've learned how to configure a hosted database and the corresponding authentication flow, let's review the next authentication option; that is, LDAP.

LDAP

Lightweight Directory Access Protocol (LDAP) can be used to synchronize user, group, and department information from an existing directory server such as **Microsoft Active Directory (AD)** to Zscaler. The user, group, and department information must be propagated from the directory service into Zscaler. End user passwords from Active Directory are never copied over to Zscaler.

During an LDAP synchronization, users are added via LDAP sync from the Directory Server to Zscaler. Users, groups, and departments that are not in Zscaler are automatically copied over from Active Directory. Users that are not in Active Directory but are in Zscaler are automatically deactivated. In summary, Zscaler will update its information to match that on the Directory Server. For redundancy, up to two forests in an organization can be synchronized. The synchronization interval can be set to daily, weekly, monthly, or manually (now).

Configuration steps

The enterprise administrator needs to log into the Admin Portal and then navigate to
**Administration -> Authentication -> Authentication Configuration -> Authentication
Settings**. Under the **Authentication Profile** tab and section, select **Active Directory**
for **User Repository Type**. Additional options appear that say **Authentication Wizard**
and **Advanced Configuration**. There is also an additional section that appears that says
Directory Synchronization. The steps involved in this configuration are outside the scope
of this book.

User authentication flow

Active Directory is synchronized with Zscaler on a periodic basis. It is a prerequisite that
must be met before any user authentication can be performed. Once that is in place, we
need to see how the authentication flow works. The end user requests a web page, and this
request is received by Zscaler. Zscaler then prompts the user for their username. The user
enters their username and then Zscaler prompts the user for their password. When the
user enters their password, Zscaler performs a BIND request over secure LDAP (LDAPS)
to Active Directory using the **distinguished name (DN)** and the password entered by the
user. User authentication succeeds if the LDAP bind succeeds. Zscaler then sends a cookie
to the end user's browser and allows the user to navigate to the initially requested web
page. User passwords are always stored and maintained on the Directory Server and are
never transferred to Zscaler.

Zscaler Authentication Bridge (ZAB)

If an enterprise's security policy prohibits it from allowing Zscaler to connect to the
organization's internal directory servers directly, or if the organization wants to bypass any
inbound firewall restrictions on its network, the enterprise can install and use a virtual
machine on its premises called the **Zscaler Authentication Bridge (ZAB)**. The ZAB then
acts as a go-between between the organization's directory servers and the Zscaler cloud.
In this case, Zscaler service talks to the ZAB, which, in turn, talks to the organization's
directory server.

The user authentication flow in this case is slightly different than in the previous case. In
this case, instead of working directly with the organization's Active Directory, Zscaler
cloud communicates with the ZAB and the ZAB works with the organization's active
directory.

Now that we've reviewed the configuration steps and authentication flow for LDAP, let's
move on to the next supported authentication method: SAML.

Security Assertion Markup Language (SAML)

In a general **Security Assertion Markup Language (SAML)** deployment model, there are three components – the **Service Provider (SP)**, the **Identity Provider (IdP)**, and the client. When this model is applied to ZIA, the service provider is Zscaler itself. The identity provider could be one of the identity management providers in the industry, such as Okta, ADFS, and so on. The client is the web browser on the end user's computer or tablet.

Let's learn about the configuration steps involved and the user authentication flow for this option.

Configuration steps

After logging into the Admin Portal, the enterprise administrator needs to navigate to **Administration -> Authentication -> Authentication Configuration -> Authentication Settings**. On the **Authentication Profile** tab and section, set **Authentication Type** to **SAML**. This enables an additional link that says **Open Identity Providers**. Clicking on that link will take the administrator to the second tab on the page, called **Identity Providers**.

One or more identity providers can be configured under this tab. The actual process for identity provider configuration and integration with Zscaler can take up an entire chapter of its own and is beyond the scope of this book!

User authentication flow

The end user requests a web page, and that request is redirected to Zscaler cloud for authentication. Once the user enters their username, they are then redirected to the configured identity provider (this step was already done by the enterprise administrator). The end user then authenticates into the identity provider with their password and if the authentication is successful, the identity provider sends a SAML assertion identity (a token) to the end user. The end user then sends this token to Zscaler and Zscaler issues an authentication token to the end user. The user is then redirected to the originally requested web page.

Now, let's look at the last authentication method available in ZIA; that is, Kerberos.

Kerberos

Kerberos is a ticket-based authentication protocol that is widely used to authenticate end users to network services. The prerequisite step is user provisioning, which is done either by using the hosted DB option or by synchronizing user data from an Active Directory or LDAP server, as explained previously. Automatic user provisioning is not supported with this option and a PAC file is needed for user authentication.

The downside of Kerberos is that it is not supported on Windows XP, Apple iOS, and Android platforms. The configuration steps for Kerberos are outside the scope of this book.

In this section, we learned the benefits of end user authentication and the various supported options by ZIA. We also looked at the steps involved in configuring each option and the end user authentication flow in each case.

Summary

In this chapter, we explored the various mechanisms we can use to forward end user web traffic to Zscaler and the benefits of user authentication. We also saw how locations can be created on the Zscaler Admin Portal for GRE and IPsec tunnels. The various user authentication options were also discussed.

Understanding these concepts allows ZIA administrators to decide on which traffic forwarding and authentication option is suitable for their enterprise.

In the next chapter, we will use this information to architect and implement a ZIA solution.

Questions

As we conclude, here is a list of questions for you to test your knowledge regarding this chapter's material. You will find the answers in the *Assessments* section of the Appendix:

1. ZIA offers multiple options for forwarding end user web traffic to Zscaler.

    a. True

    b. False

2. It is possible to enable granular, user-based policies without user authentication.

    a. True

    b. False

3. Which user authentication methods are offered by ZIA? (Select all applicable options)

    a. Hosted DB

    b. LDAP

    c. SAML

    d. Kerberos

4. For LDAP authentication, the enterprise must open its Active Directory to Zscaler and there are no alternative options.

    a. True

    b. False

Further reading

- ZIA SAML Configuration: `https://help.zscaler.com/zia/configuring-saml`

5
Architecting and Implementing Your ZIA Solution

In the previous chapters, we got introduced to the Zscaler cloud components, policy features, traffic forwarding, and user authentication options. It is time for us to look at how all these come together and create a customized ZIA solution.

The solutioning process starts with the Zscaler enterprise administrator or solutions architect asking the relevant questions to discover the current security posture of the enterprise and determine the business security requirements. The answers to these questions will decide what ZIA features will be used and how they all fit together. The solution thus developed is then implemented for a limited set of pilot end users. Any initial problems are then troubleshot, and adjustments are made accordingly. Then the solution is expanded to a larger user base, eventually covering the entire enterprise.

In this chapter, we are going to cover the following main topics:

- Analyzing the security posture of the enterprise
- Creating a customized ZIA solution for the enterprise
- Implementing the ZIA solution across the enterprise

Analyzing the security posture of the enterprise

Once an enterprise has chosen to go with the Zscaler ZIA solution, the first step is to discover the existing security posture of the enterprise. This discovery is necessary to both map the existing security requirements to ZIA features and procure the necessary hardware equipment and internet circuits necessary for the ZIA solution. Zscaler offers a free web-based question set that prompts the enterprise administrator to answer a series of questions. Let's explore this question set now.

Zscaler question set

This question set is web-based, however, be prepared to answer all the questions from start to finish in a single sitting. Unfortunately, there is no way to save the entries midway and then return to resume the remaining questions. Alternatively, the enterprise administrator can create a separate copy in a text file or a Microsoft Excel file that can be used for portability purposes. The same question set can be used to gather the answers needed for either a ZIA or a ZPA solution.

Usually, this question set is used by Zscaler Professional Services personnel who are working with an enterprise to either perform a **proof-of-concept** (**POC**) or an actual deployment. However, this can be used by anyone who is planning to deploy a Zscaler solution.

Let's now look at the various questions.

Customer Information

This section contains questions related to the enterprise that is considering Zscaler:

- **Customer Company Name** – The name of the enterprise that is considering Zscaler as their solution.
- **Customer Contact Name(s)** – The contacts who will serve as the points of contact for the enterprise.
- **Customer Contact Phone(s)** – The phone numbers of the contacts representing the enterprise.
- **Customer Contact Email(s)** – The email addresses of the enterprise personnel listed above.

- **Planned Project Kickoff Date** – This is the estimated start date for the enterprise's Zscaler project. This is helpful for Zscaler to plan for a Professional Services resource. The duration of that Zscaler resource that is available for a Zscaler deployment depends on the contract terms of the enterprise. The importance of this kick-off date is explained below.

> **Important note**
>
> Once a Zscaler Professional Service resource is assigned to a deployment project with a start date, there is no way to *"pause"* those services midway. By default, the term of the engagement runs to completion without any breaks. Depending on the contract, Zscaler *may* extend the services on a case-by-case basis, but this is not guaranteed.

- **Planned Start Date for Pilot Traffic** – Once the project is kicked off based on the date provided in the previous bullet point, the next major milestone is when the first pilot user or location is ready to start sending their web traffic over Zscaler. It is important for the enterprise to have a clear handle on how long it will take them to provision the equipment on their side, including the physical hardware, necessary software, and the configuration necessary per Zscaler pre-requisites. The enterprise should also make sure its personnel from different departments are dedicated to this project accordingly.

- **Planned Start Date for Production Traffic** – Once the pilot users or locations of the enterprise are on Zscaler, it is up to the enterprise to define the success criteria to move from the pilot phase to the production phase. If these criteria are not clear, it may drag out the pilot process for too long, jeopardizing the transition to production.

- **Planned Date for 100% Traffic Turn-up Completion** – This is the drop-dead date by which the entire enterprise needs to be on the Zscaler platform. Often, enterprises have multi-year contracts with security vendors. If the migration to the Zscaler platform does not happen by the contract expiration date of the current security vendor, the enterprise may be left in an awkward position with security gaps. It is often expensive for the enterprise to renew the contract with the current security vendor for several months just because of a gap of a few days. Hence, it is always recommended to keep a buffer time between those two dates (the 100% transition to the Zscaler platform and the current security vendor contract expiration date).

- **Deployment Objectives** – This is a free-flowing optional text field that can contain the main reasons the enterprise has chosen Zscaler. Perhaps the primary reason is security against malware, or security for the mobile users, or if an enterprise does a lot of backups, sandboxing may be its primary objective. Although an optional field, noting these objectives down allows the enterprise administrator to prioritize those features applicable to these objectives.

- **Your Name** – The name of the person who is filling out this question set. This person may be a single point of contact between the enterprise and Zscaler.

- **Your Phone** – The phone number of the person completing this question set.

- **Zscaler Zendesk Ticket # (if available)** – Often, Zscaler Professional Services also offers a **Project Manager (PM)** who drives the project from start to finish. They create a Zendesk (the ticketing system used by Zscaler) ticket and document the project's notes and progress against this ticket. This is an optional field.

Click **Next** to navigate to the next page.

Products Purchased

As mentioned earlier, this question set can be used to gather answers to deploy either a ZIA or a ZPA solution. Zscaler Shift is out of scope for this book. Select the checkbox against **Zscaler Internet Access (ZIA)** and click **Next** to navigate to the next page. The **Zscaler Private Access (ZPA)** option will be covered in detail in *Chapter 7, Introducing ZTNA with Zscaler Private Access (ZPA)*.

Approx. Number of Users Planned to be on Zscaler

Both Zscaler Professional Services and any third-party **managed security services provider (MSSP)** usually estimate the scale of the project deployment based on this number. If there are unauthenticated users using ZIA, for those transactions, Zscaler counts about 250 web transactions per day as a user:

- **Approx. Number of Total Production Users (Including Guests & Contractors)?** – This number should include the total number of users across the enterprise. If service accounts will be used by endpoints that will also perform authentication and will have granular policies applied against them, they should be included as a user in the count.

- **Approx. Number of Remote Users (Road Warriors)?** – Zscaler calls the users who are mostly mobile (such as working from home, or who travel a lot) Road Warriors. The reason for a separate count of mobile users is specifically from the perspective and need of the ZCC application.

Click **Next** to navigate to the next page.

Scope – Traffic Sources

This section has the various types of sources for the end user traffic:

- **Managed Users on Net** – These are the employees who usually work in a corporate office location. Some of these users can occasionally work from home, or from a remote location such as an airport or a hotel.

- **Managed Users Off Net (Road Warriors)** – These are the employees that either work offsite from home or travel extensively. They can also occasionally come to the office as mentioned in the previous case.

- **Guest Wi-Fi Users** – These are the guests such as external vendors, contractors, or customers who need temporary internet access.

- **BYOD Users** – **Bring Your Own Device (BYOD)** is a popular option available nowadays to many employees. In this case, the employee brings and uses a personally owned device and can use it for work purposes using an isolated or virtualized environment.

- **Servers** – These are either physical or virtual machines that serve a specific purpose for the enterprise. Some examples are web servers, **File Transfer Protocol (FTP)** servers, and so on.

- **Virtual Desktop Infrastructure (VDI) Users** – Instead of handing out physical computers or tablet devices, some enterprises allocate a Virtual Desktop on a cloud platform that end users can access through any generic computer. Such end users are called VDI users.

- **Kiosk/Shared Machines** – Kiosks are machines that are usually provisioned by an enterprise to provide self-service to customers. These can be seen at airports and train stations where a passenger can log in using their passport or driver's license number to check in and print their boarding pass or a train ticket. Shared machines are computers that generally stay in the same location and are used by multiple end users on a shared basis, typically in a 24 x 7 call center or support desk environment.

- **IoT (Connected Devices)** – **Internet of Things (IoT)** devices are typically devices that directly interface with the internet. A few examples are heat sensors, thermostats, and shipping containers.

- **Android Devices** – These are computers, tablets, and smartphones that run the Android operating system.

- **iOS Devices** – These are computers, tablets, and smartphones that run Apple's iOS operating system.

- **Other** – This is a free-flowing textbox that can be used to capture any traffic sources that have not been covered by previous categories.

Click **Next** to move forward with this process.

Package Management

In a medium-to-large enterprise environment, typically, end users are not allowed to download software from unauthorized sources such as the internet. Instead, the internal IT department first obtains the application software packages from an authorized source such as the software manufacturer, tests those packages, and optionally customizes them for the enterprise, and then distributes it to the employees of the enterprise. This end-to-end process is called **package management**:

- **How are Application Packages Deployed for the Following Managed Devices?** – The options presented here are for various operating systems and the corresponding package management choices. The default option for each platform is **To Be Determined (TBD)** and the very next option after that is **Operating System (OS) Not In use**. Let's examine some of the choices other than these two for each platform.

- **Windows – System Center Configuration Manager (SCCM) / Group Policy Object (GPO)** are very commonly used by enterprises for their users using the Microsoft Windows platform. There are several third-party package management options also available.

The other platforms that appear on this page are **Mac OS**, **GNU/Linux**, **iOS**, and **Android** devices, and **Chrome OS**. There is a free-flowing text field at the very end that can be used to capture any other package management options not already listed. Click **Next** to continue with this discovery process.

Antivirus/Security Solutions

Like the previous option, this section asks what antivirus security solution is in use within the enterprise. This information is useful to know, especially for the ZCC app because certain antivirus or security solutions could interfere with the operation of the ZCC app.

Which Antivirus/Security Solutions Are in use for the Following Managed Devices? – Under this question the operating system platforms are listed again, except Chrome OS. The listed platforms are Windows, Mac OS, GNU/Linux, iOS, and Android devices. The free-flowing text field at the end can again be used to capture unlisted security solutions. Click **Next** to move to the next page.

Network Segmentation

Network segmentation is a best practice where the entire network of the enterprise is logically divided into various segments, each segment serving a specific purpose. These segments are then used to determine the trust direction of the traffic. For example, let's say you have network segments called guest Wi-Fi network, employee network, and company servers. It might then be a good idea to block guest Wi-Fi networks from accessing the company server segment but allow the employee network to access the company servers. Without a proper network segmentation approach, it can become very confusing to implement an enterprise security solution:

- **How is the Network Segmented for the Following Traffic Sources?** – This question is asking if the network is segmented for the various traffic sources that were already mentioned earlier. Here, the traffic source options are **Guest Wi-Fi**, **BYOD**, **IoT**, **Remote Desktop Services**, **Application Servers**, and **Kiosk/Shared Machines**.

The choices under each of the traffic source options mentioned in the preceding are as follows:

- **Segmented at all Locations (best practice)** – As mentioned previously, it is a best practice to segment the network at all the applicable enterprise corporate locations.

- **Not Segmented at any Locations** – This option is self-explanatory. If this option is selected by the enterprise administrator who is filling out this page, the question of why the network is not segmented should be raised with the proper departmental teams. This makes sure that such a critical issue is addressed well before the start of a Zscaler deployment or migration.

- **Segmented for Some Locations** – Again, this should serve as a wake-up call for the enterprise network team who can go back to the drawing board and make sure all locations have transitioned to a proper and consistent network segmentation approach.

- **Not Used/Deployed** – This choice could probably be applicable for a very small enterprise that has a handful of users and a flat network or an internet-based company that does not have a corporate location.

Internet Egress Points

Each corporate office location typically has a local internet connection or all the corporate locations route traffic to a central data center that has one or more large internet egress points through a corporate firewall. This section attempts to capture that information. Let's look at the non-default TBD options:

- **How is Internet Traffic Routed?** – The first choice is **All Branches Backhaul to Hub Locations** (hub and spoke model). We discussed this scenario at length in *Chapter 1, Security for the Modern Enterprise with Zscaler*. The second choice is **Branches have Local Internet** (breakouts). Due to the low-cost availability of internet access, many enterprise corporate locations are moving from an expensive MPLS type of connection to either a single or a dual internet connection. This choice describes such an option. Another popular option along with a dual internet connection is **Software-Defined Wide Area Network (SD-WAN)**, which is out of scope for this book. The third and last choice is **Some Breakout and Some Backhaul to Hubs**. This choice can be chosen if the enterprise has a combination of centrally-located data centers as well as regional locations with local internet breakout circuits.

- **Is there a Default Route on the Local Network?** – The choices are either yes or no.

- **Which Statement Below Best Describes Current Internet Bound Throughput from Each of the Locations?** – The choices are **All Locations < 200 Mbps, Some Locations > 200 Mbps**, and **Some Locations > 1 Gbps**. The thought behind this question is the determination of the current throughput requirements and planning for any increase in throughput of the internet traffic based on the growth of the enterprise in the future. We do not want to design a Zscaler solution where we quickly hit a throughput limit that causes us to revisit the deployed solution frequently.

- **Total Number of Locations with Internet Egress Points?** – This is a free-flowing text field because this number varies for each enterprise. If you have a single location with more than one internet egress point, then count the number of internet circuits at the single location rather than counting just that one location.

Internet Egress Points – Geographic Locations

Although the Zscaler cloud is present and available in various countries and geographies, this section attempts to capture the footprint of the enterprise, especially if the enterprise is a multinational with a presence in more than one country or geography. If an enterprise location does not have an in-country Zscaler data center available, then the enterprise administrator needs to be prepared to connect into a different, neighboring country's Zscaler data center:

- **Please Select the Geographic Locations where you have Internet Egress Points –** The choices are textboxes with the following locations: **USA**, **Canada**, **Other North America**, **Europe**, **South America**, **Africa**, **China**, **Middle East**, and **Other APAC**. There is also a free-flowing text field to capture any special cases. Click **Next** to continue with the process.

The next page says **Please Provide Info About Your Top 5 Internet Egress Points**. This is to capture the internet speeds of the top five egress points. The fields for each egress point are **Location/City/State/Country**, **Approximate Number of Users**, **Approximate Peak Download Bandwidth**, **Approximate Peak Upload Bandwidth**, and **Provide the Static IP(s) for this Location**. The intent here is to understand how many GRE or IPsec tunnels will be needed for each internet egress point. Once you are done adding additional egress points by clicking on the **Add Another** button, click on the **Next** button to proceed to the next step.

GRE Infrastructure

This section attempts to understand if there is a GRE-capable device available at the enterprise location. Let's explore the non-TBD options on this page:

- **Is there a GRE Capable Device Available?** – The first option is **Available for All Locations (Best Practice)**. GRE tunnels are the preferred option for a Zscaler deployment. This is both due to the simplicity and higher throughput offered by the GRE tunnels when compared to IPsec tunnels. The second option is **Available for Some Locations**. The thought here is that if GRE-capable devices are not available for some locations, the enterprise administrator may want to start procuring those necessary devices or look for alternative traffic forwarding options for those locations. The third and last option is **Not Available for any Locations**.

- **Do GRE Devices Support Automatic Tunnel Failover?** – The first option is **Yes for All Locations (Best Practice)**. As discussed during traffic forwarding, it is best practice to configure at least two tunnels to Zscaler (at least two for redundancy, although a single tunnel also works) on each enterprise device in a primary/ secondary setup. If the devices support automatic tunnel failover, it increases the availability of the service without the need for manual intervention. The second option is **Yes for Some Locations**. For locations that do not support automatic tunnel failover, the enterprise needs to accept potentially longer downtimes and the need for a manual tunnel failover requiring human intervention. The third and last option is **No for All Locations**.

- **Will the GRE Device Send Internal IPs (UnNatted) to Zscaler?** – The three options are the same as the previous case, namely, **Yes for All Locations**, **Yes for Some Locations**, and **No for All Locations**. Most GRE-capable devices send traffic to the tunnel using policy-based routing, thus preserving the internal client IP addresses. However, if the GRE-capable device performs **network address translation (NAT)** before it sends traffic through the tunnel, the enterprise should consider disabling NAT to allow Zscaler to view the internal IP addresses. This enables Zscaler to use those internal IP addresses for logging and reporting purposes. Click **Next** to proceed to the next step.

GRE Vendors

This section gathers the vendor name(s) of the GRE capable devices managed by the enterprise. The checkbox options are **Cisco**, **Juniper**, **F5**, **Barracuda**, **HP**, **Fortinet**, **Palo Alto**, and **None**. There is also an **Other** field that can accept free-flowing text if the preceding list does not capture the vendor.

IPSEC Infrastructure

Like the *GRE Infrastructure* section, this section addresses the availability question of the IPsec capable devices. Let's review the only additional option here:

- **Does the IPSEC Device Support Null Encryption?** – Null encryption refers to the choice of not using encryption in a system where various encryption choices are offered. When this option is used, the original text remains the same even after encryption, which can be utilized for testing or debugging purposes. The three options here are **Yes for All Locations (Best Practice)**, **Yes for Some Locations**, and **No for All Locations**. Click **Next** to continue.

IPSEC Vendors

This page collects the vendor name(s) of the IPsec-capable devices that will be used by the enterprise. The list of vendors is a little bit different from the GRE vendors. The options here are checkboxes for **Cisco, Juniper, Palo Alto, Checkpoint, SonicWall, Barracuda, HP, Fortinet, F5, Cisco Meraki**, and **None**. If these choices do not apply, the **Other** field can be used. Click **Next** to proceed to the next step.

Preferred Traffic Forwarding Methods

Recall the various traffic forwarding methods discussed in the previous chapter. This section enumerates the various traffic forwarding methods and certain preferred combinations:

- **Preferred Traffic Forwarding Method for Corp Locations** – There are several choices under this option. The first and preferred choice is **GRE + ZCC App (Best Practice)**. The remaining choices are **GRE + PAC, IPSEC + PAC, IPSEC + ZCC App, GRE Only, IPSEC Only, PAC Only, ZCC App Only (recommended for locations less than 300 Mbps), Proxy Chaining, Explicit Proxy**, and **None**.

- **Preferred Traffic Forwarding Method for Road Warriors** – There are only two choices for remote users who mostly work offsite. The first and preferred choice is **ZCC App (Best Practice)** and the second choice is using **PAC**.

- **Preferred Traffic Forwarding Ports** – The first and preferred choice is to send **Any TCP/UDP/ICMP (Best Practice)** port traffic and the second choice is to **Send Only Web Traffic on ports 80 and 443**. Click **Next** to continue with the process.

Existing Secure Web Gateway

If the enterprise is deploying Zscaler ZIA for the first time, this section does not apply. However, in many cases, an enterprise might be migrating from a current secure web gateway solution to Zscaler. It is important to know if this is the case. If it is indeed a migration, the enterprise administrator may want to map the current secure web gateway policy to Zscaler. Often, this is not a straightforward process because the feature set and behavior of one vendor differs significantly from Zscaler.

Instead of attempting a one-to-one policy and feature migration, the enterprise administrator should take this opportunity to clean up unused and outdated policies and consolidate several similar policies before migrating to Zscaler. This is often a very labor-intensive operation and enough resources should be allocated to that effort.

It may not be possible to perform a one-to-one mapping from an existing secure web gateway product to Zscaler and the enterprise may be disappointed if it has that expectation. On this page, select the appropriate dropdown under **Please Select the Secure Web Gateway Currently in Use**. The **Other** field can be used if a certain secure web gateway is not found in the drop-down list. Click **Next** to carry on with the process.

Traffic Forwarding for the Existing Proxy

This section tries to understand how the end user traffic is being forwarded to the existing proxy identified in the previous step:

- **How is End User Traffic Forwarded to the Existing Proxy?** – The choices against the checkboxes are **PAC File**, **Proxy Chaining**, **Registry**, **Agent**, **L2/L3 Forwarding**, **WCCP**, **Browser Proxy Configuration**, **TBD**, and **Other**. Click **Next** to proceed with the discovery process.

Networking – Interop

This page captures the existing general setup of the networking within the enterprise:

- **Restriction on Source IPs** – The first option is **Yes; Source IP Preservation is needed for some cloud apps**. Recall this option when we discussed the **Virtual Service Edge (VSE)** in *Chapter 2, Understanding the Modular Zscaler Architecture*. Sometimes, an enterprise accesses certain third-party vendor sites and those third-party vendors whitelist the IP address range of the enterprise. If the enterprise moves to Zscaler, then the end users will access the third-party vendor sites using the Zscaler IP address range. The third-party vendor may not be thrilled to whitelist the entire Zscaler IP address range. If that is the case, then the enterprise may have to deploy a Virtual Service Edge that preserves the source IP of the enterprise. The second option is **No, Source IP Preservation is not needed for any cloud apps**. If this option is chosen by the enterprise based on its business needs, then there is no additional work to be done such as deploying and maintaining the VSEs.

- **Any Applications using Application Layer Gateway (ALG) Feature on FW? (ex. VoIP, Active FTP)** – The choices for this option are **Yes** or **No**. The significance of this option lies in the fact that Zscaler does not support VoIP protocols such as **Session Initiation Protocol (SIP)**, H.323, H.248, **Voice over Internet Protocol (VoIP)**, and **Active File Transfer Protocol (FTP)**. Zscaler recommends the enterprise administrator bypass these protocols from Zscaler by adjusting the configuration on the firewall or the router of the enterprise when configuring the GRE or IPsec tunnel. Zscaler supports passive mode FTP.

- **Is WinHTTP Configuration Deployed?** – Microsoft **Windows HTTP Services (WinHTTP)** provides developers with an HTTP client **Application Programming Interface (API)** to send requests through the HTTP protocol to other HTTP servers. The choices are **Yes** or **No**.

- **Is WPAD Deployed?** – The **Web Proxy Auto-Discovery (WPAD)** protocol is used by end user clients to locate the **Uniform Resource Locator (URL)** of a configuration file using DHCP and/or DNS discovery methods. Once the configuration file is detected and downloaded, it can be executed to determine the proxy for a specific URL.

- **Select Cloud Access Security Brokers (CASB) Vendor** – The choices in the drop-down list can be selected if the enterprise uses a CASB as part of its security solution. If the choices do not cover the vendor, the free-flowing **Other** text field can be utilized.

Click **Next** to proceed to the next page.

Remote Access VPN Deployment

Recall that there is a split-tunnel mode and a full-tunnel mode when the ZCC App is used along with a VPN client. This section captures some of the details related to the existing VPN deployment of the enterprise. These details will help with how the VPN client interacts with the ZCC App:

- **Which VPN Clients are Supported in your Environment?** – The various vendor choices are listed in the dropdown and the appropriate one can be selected.

- **How is Remote Access VPN Connection Established?** – The three choices available here are **On Demand**, **Always On**, and **Not Applicable**.

- **How is Remote Access VPN Deployed?** – The possible choices are **Split Tunnel (Best Practice)**, **Split Tunnel** and **Split DNS**, **Full Tunnel**, and **Not Deployed**.

- **Can the Internal DNS Servers Resolve External Public Domains?** – The choices are **Yes** or **No**. This becomes especially important when PAC files are hosted on the Zscaler cloud and the end users will need to resolve external public domains using the internal DNS servers.

Click **Next** to continue the process.

Authentication for ZIA

This page checks to see whether the enterprise has a business requirement to identify the end users who are accessing the internet. Recall that to identify end users and apply user-based policies, authentication is necessary:

- **Do you need to Identify Users Accessing the Internet?** The first option is **No, Do not Require** and the second most used option is **Yes, To Configure User/Group Specific Policies and Reporting**. Click **Next** to proceed to the next section.

User Management

This section gathers the details on what end user authentication options are available within the enterprise:

- **Please Select Current SAML 2.0 Federation Vendor** – This option has a dropdown with several choices for the federation vendor. Zscaler integrates with several leading SAML 2.0 federation vendors.

- **Can SAML 2.0 Federation Authenticate all Required Users?** – The first choice is **Yes**, in which case, this authentication can be used for all users. If the choice of **No** is selected, then the enterprise administrator may have to use an alternate authentication method or not use authentication at all for the unsupported users.

- **Is SAML 2.0 Federation Accessible Over the Internet for Remote Users?** – The choices are **Yes** or **No**. This is especially important to know for the remote users using the ZCC App.

- **Is Integrated Windows Authentication (IWA) in Use?** – IWA enables end users to log in with their Windows credentials with the client sending the credentials in an Authorization header. This is best suited for an intranet environment. The first option is **In Use** and the second option is **Not Used**.

- **What is your User Directory Type?** – The choices available in the drop-down list are **Active Directory**, **OpenLDAP**, and **Other**.

- **How many Active Directory Forests are there?** – The available choices in the drop-down list are **one**, **two**, **three or more**, and **None (AD Not in Use)**.

If needed, use the **Other** free-flowing text field to make any relevant notes not covered already by the options.

Click **Next** to continue with the discovery process.

Authentication and Provisioning Options

This section asks more questions that are helpful in answering which authentication and provisioning methods can be chosen:

- **Can you allow inbound firewall (FW) rules from Zscaler to your Directory?** – The only valid choices are **Yes** or **No**. Recall that if an enterprise does not want to open inbound firewall rules from Zscaler to its internal Active Directory due to security or business needs policy, then it can make use of the **Zscaler Authentication Bridge (ZAB)**.

- **Do you have ESXi Infrastructure?** – The choices available here are again **Yes** or **No**. If the enterprise wants to use a ZAB, it needs to have the VMWare ESXi infrastructure to run the virtual machine.

- **Do you Require Policies Applying to Nested Groups?** – The choices are **Yes** or **No**. Remember that Zscaler does not support nested groups. The enterprise administrator must manually identify specific group names within each nested group and add them to the **Group** filter.

- **Do you Require Single Sign On (SSO)?** SSO is an authentication method where an end user logs in with a single ID and password and uses that to log in to several different systems. The thought process here is that the enterprise administrator or users can authenticate against their **Identity Provider (IdP)** and then log in to the Zscaler service using SSO.

Click **Next** to proceed.

Policy

Recall all the policy options that were explained in *Chapter 3*, *Delving into ZIA Policy Features*. This page tries to understand the policy requirements of the enterprise:

- **Do you plan to Inspect SSL Traffic?** – There are three available choices. If the enterprise decides to inspect SSL traffic, the two sub-choices available are **Yes, Using Custom Root Certs (Best Practice)** or **Yes, Using Zscaler Root Certs**. Recall the differences between using a custom enterprise-created certificate versus using a Zscaler root certificate under the *SSL Inspection* section of *Chapter 3*, *Delving into ZIA Policy Features*. The third and last choice is **No**, which means the enterprise will not inspect SSL traffic.

- **Do you plan to use Bandwidth Controls?** – If the enterprise has limited bandwidth and chooses to use bandwidth controls to save bandwidth for business-critical apps, there are two choices. The first one is **Yes, Using Zscaler Feature** and the second option is **Yes, Using 3rd Party Solutions**. The third and last choice is **No**.

- **Do you plan to use Zscaler DLP Feature?** – If the enterprise chooses to use the Zscaler DLP feature, there are two sub-options and they are **Yes, With ICAP Server Integration** and **Yes, With No ICAP Server Integration**. The third and last option is **No, Zscaler DLP Feature Will Not Be Used**.

- **Do you plan to use Cloud Sandbox?** – If the enterprise wants to use the cloud sandbox for unknown and potentially malicious signatures, the options are **Yes, Using Zscaler** and **Yes, Using 3rd Party Solutions**. The third and last option is **No**.

- **Do you have an Existing Security Appliance In-Line with a Default Block Policy?** – The options are **Yes** or **No**. Remember that Zscaler offers a Cloud IPS and this question is intended to be the answer for the enterprise that does not have an existing IPS.

Click **Next** to continue.

Logging

This page tries to understand the logging requirements of the enterprise:

- **Do you plan to Integrate Zscaler with a SIEM System?** – The choices in the drop-down list are **Yes** or **No**. Recall that the NSS VMs could be used to integrate with an SIEM managed by the enterprise.

- **What are the Log Requirements on Zscaler Nanolog Servers?** – The first choice is **6 months (Default)** and the second choice is **1 Year (Requires Additional Purchase + Approval)**. Zscaler stores the logs on its Nanolog Servers for up to 6 months by default. Any additional log storage beyond that timeframe is possible by purchasing it from Zscaler for up to a year or integrate with the enterprise SIEM and storing them there for as long as needed.

- **Do you need to Comply with EU Data Protection Regulations?** – The answer choices are **Yes** or **No**. If an enterprise has a footprint in the **European Union (EU)**, the log activity for end users in the EU should be stored only within the EU region. This effectively tells Zscaler to be extra careful when handling the EU user activity logs.

- **Do you Require all logs to be Stored only within United States?** – Again, the answer choices are **Yes** or **No**. If there is a requirement per US government regulation or if the enterprise has sensitive information such as a defense contractor, it may be bound by regulation that calls for the user log activity to be stored within the US. This is like the EU case seen previously.

Click **Next** to proceed to the next section.

SIEM System

Please Select Current SIEM System – The enterprise administrator can choose the SIEM system that is currently in use by the enterprise. This will help decide whether the current SIEM system can work with Zscaler.

This is the last page of the question set. Click on the **Submit Design Questionnaire** link to email the saved content to all the email addresses that were entered on the *Customer Information* section page.

Note that this question set should serve as a logical flowchart for the enterprise administrator. At the same time, it should give all the necessary information required to architect a customized ZIA solution. Let's now review the important high-level steps involved in the creation of the customized ZIA solution.

Creating a customized ZIA solution for the enterprise

The Zscaler question set explored in detail so far should be completed by the enterprise Zscaler administrator with the help of various inter-departmental teams of the enterprise. This includes technical teams such as networking, firewall, IT, and non-technical teams such as project management. This discovery process takes the longest, but it is the most critical step because it lays the foundation for future steps.

The high-level components of a customized ZIA solution are traffic forwarding, user authentication, and policy. Let's explore the first component in more detail.

Traffic forwarding

The first component of a ZIA solution is traffic forwarding. There are so many ways end user traffic can reach Zscaler. Here are some common scenarios.

Internet start-up or small businesses

Let's look at the case of an internet-based start-up company or a small business with very few local or regional locations. An internet-based start-up company will probably not have a fixed office location where employees go to work. Most of them will probably work remotely from home or while traveling using the internet. A small business with a few local and regional locations will probably not have many employees at each location. At the same time, their network equipment and personnel budget may be the bare minimum. So, it does not make sense for companies like these to invest in a lot of physical hardware. The internet should be leveraged for forwarding the end user traffic to Zscaler.

The use of the ZCC App is best suited for this scenario. It works on the most common operating system platforms, including computers, smartphones, and tablets. The company owner or administrator can easily distribute the link to download the ZCC App to its end users. The administrator can also enable the end users to automatically update the ZCC App.

Medium-sized enterprise

A medium-sized company may have a few hundred employees with a few spread-out corporate locations, usually within the same country. Each corporate location may have tens of employees who also work remotely on an occasional basis. A mix of GRE or IPsec tunnels and the ZCC App is best suited for this scenario to send end user traffic to Zscaler.

Large multinational enterprise

The last scenario is that of a large enterprise. Such companies could have their footprint within a single country or in several countries spanning different continents. Typically, they have a large headquarters location, a central data center or a few regional data centers, and several regional corporate offices.

GRE or IPsec tunnels with PAC files are the preferred method for end user traffic forwarding in this scenario. The data centers usually have redundancy with respect to geographically dispersed locations, network equipment, and internet connections. The ZCC App can also be leveraged for employees who travel quite a bit and for employees who work out of hours such as production support and installation, which does not conform to the standard business work hours from a corporate location.

Let's now look at the user authentication options.

User authentication

Recall that user authentication is necessary to identify a user with a web transaction and to apply granular policies at the user level, no matter where the user logs in from.

Internet start-up or small business

Because of the small number of users and lack of budget for a dedicated Zscaler administrator, it is often convenient to use the hosted database option in this situation. It is very easy to manage as everything related to user authentication can be performed on the Zscaler Admin Portal, which serves as a one-stop-shop.

Medium-sized enterprise

Many medium-sized enterprises typically use an on-premises Active Directory, although cloud-based identity management solutions are rapidly gaining in popularity. LDAP would be the recommended choice for user authentication in this scenario.

Large multinational enterprise

Several large legacy enterprises use a mix of on-premises Active Directory along with cloud-based technologies. For automatic economies of scale, we recommend cloud-based identity management solutions. Let's explore the last major component, policy.

Policy

Policy is a set of rules governing end users using the enterprise assets to access resources on the internet. Let's look at the policy recommendations.

Internet start-up or small business

As mentioned earlier, it is very much possible for the business owner to serve as the Zscaler administrator. Hence, it is probably best for the small business to use the Zscaler default recommended policy in this scenario.

Medium-sized enterprise

It should be possible for the Zscaler enterprise administrator to perform a quick discovery of the business security needs and develop a policy based on the Zscaler default recommended policy using Zscaler root certificates for SSL inspection.

Large multinational enterprise

Significant time needs to be invested for the discovery of the business security needs for the large enterprise. Because it operates in so many countries and continents, a one-size-fits-all approach will not work in this scenario. The Zscaler default recommended policy can be used as a starting point, but the effective policy needs to consider the business requirements from various teams.

Let's now look at how we can implement this customized ZIA solution.

Implementing the ZIA solution across the enterprise

Implementing the customized ZIA solution consists of Planning, Configuration, Pilot, and Production Rollout stages. Let's start by looking at the Planning stage.

Planning

The first stage of implementing a ZIA solution is Planning. Once the enterprise evaluates the various secure web gateway solutions in the market and chooses Zscaler (preferably after a proof of concept), the planning phase begins.

Once the contract is signed with Zscaler, the enterprise should dedicate one or a few resources to serve in the capacity of Zscaler **enterprise administrator (EA)**. The EA is responsible for contacting the enterprise project sponsor and getting the list of all the names of resources from various departments to assist with the project. A **Project Manager (PM)** should also be assigned.

The enterprise PM contacts the Zscaler PM and account team. A kick-off call is then arranged where all the parties introduce themselves. Recurring meetings are established and during the first few weeks, the Zscaler question set is completed for the ZIA product. During this discovery process, the necessary physical equipment, internet circuits, virtual machine infrastructure, and so on are procured. A design document is prepared during this phase that is reviewed by all parties in terms of its viability and then approved by the enterprise project sponsor.

We cannot emphasize the importance of the planning phase as it sets the direction for the next stage, which is Configuration.

Configuration

Based on the approved design plan, equipment is installed and configured by the enterprise personnel. The EA performs the configuration steps needed on the Zscaler Admin Portal and engages the Zscaler provisioning team via the ticketing system for the creation of locations, and so on. The EA creates test user accounts that can be used to verify the traffic forwarding, user authentication, and the application of Zscaler policy features chosen during the Planning stage.

It is a good idea to test the traffic failover feature between primary and secondary: Zscaler tunnels, GRE or IPsec devices, internet circuits, data centers, and so on. This is done during this stage because there are no pilot or production users using the ZIA solution yet.

During this stage, friendly pilot users and/or locations are identified by the EA in association with the enterprise project sponsor. The project now moves to the next phase, the pilot rollout.

Pilot rollout

During this phase, the pilot users are provisioned into the Zscaler Admin portal so they can authenticate using the mechanism decided upon during the planning phase. The users are then switched to the Zscaler ZIA solution and enough time is allocated so that the pilot users may test all their business applications and tools.

These pilot users report any issues encountered using the ZIA solution to the EA. The EA then works with Zscaler and the internal departments as applicable to resolve those issues or find workarounds. At the end of the pilot testing phase, the pilot users should have completed testing most of their business applications.

This leads us into the next phase, which is the production rollout.

Production rollout

Production changes within an enterprise usually follow strict guidelines to minimize business impact. A maintenance window is obtained by the EA by following the enterprise change management procedures. This window is communicated to all the stakeholders well in advance and is usually scheduled outside of business hours.

During the change window, the already configured equipment is brought online and a larger population of end users are switched to the Zscaler solution path. The end user experience is monitored for a few weeks and any resulting issues are addressed by the EA in association with Zscaler and internal departments as applicable.

The end user and/or location count will then constantly be increased as the production rollout expands, eventually encompassing the entire enterprise. Once this is completed, the project is officially considered complete.

Summary

In this chapter, we looked at how to architect and implement a customized Zscaler ZIA solution. The critical component is asking the questions recommended by the Zscaler question set. The enterprise then assigns a Zscaler enterprise administrator who orchestrates various internal company departments and Zscaler and proceeds to implement the ZIA solution.

You have now learned what questions to ask of your enterprise teams when architecting a customized ZIA solution for your enterprise. It also helps you in planning the various stages of a successful ZIA deployment.

In the next chapter, we will look at how this customized ZIA solution can be optimized for best performance and how to troubleshoot issues commonly seen by end users.

Questions

As we conclude, here is a list of questions for you to test your knowledge regarding this chapter's material. You will find the answers in the *Assessments* section of the *Appendix*:

1. It is recommended to use the Zscaler question set while designing a custom Zscaler solution.

    a. True

    b. False

2. Implementing a Zscaler solution involves the following steps. Select all that apply.

    a. Configuration

    b. Production Rollout

    c. Pilot Rollout

    d. Planning

    e. None of the above

3. Production changes can usually be performed during business hours without an approved change window.

    a. True

    b. False

4. An enterprise needs to have a VM infrastructure to implement a ZAB or a VSE.

    a. True

    b. False

Further reading

- Zscaler question set: `https://zingtree.com/host.php?style=panels&tree_id=341454960`

- Zscaler Shift: `https://help.zscaler.com/shift/about-shift`

- Zscaler SD-WAN Security Solutions: `https://www.zscaler.com/solutions/sd-wan-security`

6
Troubleshooting and Optimizing Your ZIA Solution

After learning how to architect a custom **Zscaler Internet Access** (**ZIA**) solution, it is time to put that solution into a day-to-day operation. To make the most out of the deployed ZIA solution, the enterprise administrator needs to take care of a few aspects.

Anyone who has been in any type of steady-state operation will almost immediately tell you that it involves working with trouble reports initiated either proactively created by a network monitoring tool or reactive tickets created by end users over the phone or through a web portal.

These trouble tickets are then routed to a generic help desk, where the associates need to know how to identify a potential Zscaler problem and engage the proper points of contact. For this reason, there needs to be a comprehensive and standardized troubleshooting process documented for the help desk. This documentation needs to include the various points of contact within the enterprise, such as the firewall team, the network team, IT operations, and the proxy team. The troubleshooting process also needs to have one or more customized flows described for each type of problem. Otherwise, it might create a nuisance if everyone is engaged for every problem.

When the trouble is finally escalated to the appropriate Zscaler enterprise administrator, the administrator will have to start at a baseline. The dashboards offered by the ZIA Admin Portal is the first place the administrator should reference. Custom widgets can be created by the administrator to suit the enterprise's needs. Streamlining this entire end-to-end operations process enables the enterprise to provide a great end user experience while saving on costs.

In this chapter, we are going to cover the following main topics:

- Setting up proactive ticketing and alerts

- Producing reports for management review

- Generating custom widgets for the ZIA Dashboard

- Creating a unified ZIA troubleshooting guide

Technical requirements

Knowledge of operating system commands, such as command prompt, terminal, ping, and traceroute, is helpful to gain an understanding of the content in this chapter.

Setting up proactive ticketing and alerts

We have already mentioned that reactive tickets are usually created when end users call the help desk or via an online ticketing portal. In this section, we will look at the various ticketing and alerting options available with ZIA. First, let's begin with the native alerting mechanism offered by the ZIA Admin Portal.

ZIA alerts

The enterprise administrator can log in to the Admin Portal and then navigate to **Administration** > **Alerts**. On the first tab, **Define Alerts**, click on the + icon to add a new alert definition. The resulting pop-up window has the following options:

- **Status**: Alerts can be in an **Enabled** or a **Disabled** state. As a best practice, it is good to start with a new alert definition in a disabled state until the administrator has had a chance to fine-tune the settings.

- **Alert Name**: The administrator can select the specific event of interest from the drop-down menu list.

- **Alert Class**: Based on the previous selection, the value for this field is automatically set to the appropriate alert class. Those classes are **Security**, **Access Control**, **System**, **Comply**, and **Patient 0**.

- **Minimum Occurrences**: This sets the number of minimum occurrences this alert occurs. The available values are **1, 5, 10, 100**, and **1000**.

- **Within Time Interval**: This field is used in conjunction with the previous field. The logic is if the X number of minimum occurrences happens within the Y time interval, then trigger this alert. The available values are **5 minutes**, **15 minutes**, **30 minutes**, **one hour**, and **one day**.

- **Applies To**: By default, the option is set to **Organization**. That means, if the X number of minimum occurrences occur within the Y time interval across the entire organization, then trigger this alert. The other options are for a **Location**, a **Department**, or a **User**.

- **Severity**: This indicates the severity level of this alert. This level can be chosen by the enterprise based on its business needs. The available values are **Critical**, **Major**, **Minor**, **Info**, and **Debug**.

On the second tab, **Publish Alerts**, the alert defined just now can be sent to the interested subscribing parties. Click on the + icon to **Add Alert Subscription**. The pop-up window that appears has the following options:

- **Email**: Enter the email address of an individual or a group e-mail address for the enterprise help desk, or security team, for example.

- **Description**: This is a free-flowing text field where the purpose of this alert subscription can be defined.

Under each of the types of alert class, select the severity level that this subscribing party will receive. For example, the Tier 1 help desk can subscribe to the **Minor** level of alerts and the dedicated security team can subscribe to the **Critical** alerts. This way, the alert levels can be segmented based on the subscribing party.

Under the third and final tab, **Global Configuration**, the frequency of the alerts can be set. For the **Resend Active Alerts Every** option, the drop-down choices are **30 minutes**, **1 Hour**, **6 Hours**, **12 Hours**, and **24 Hours**. The administrator can choose the right option based on the security needs of the enterprise.

ZIA ticketing

The Admin Portal does not offer any native ticketing tools. If the organization has a need for proactive ticketing, they need to set up their SIEMs to receive the real-time web and firewall logs from Zscaler NSS virtual machines, and then build the event correlation and ticket generation logic into their SIEM and integrate it with their preferred ticketing platform.

The ZIA Admin Portal offers several types of NSS feeds that can be consumed by the SIEM:

- **Web Logs**: The administrator can configure up to eight NSS feeds for the web logs that are generated as and when the end users browse the web through ZIA.

- **Firewall Logs**: If the enterprise has chosen to use the ZIA Firewall, then the firewall logs can be sent to the SIEM. In the same way as web logs, up to eight NSS feeds are supported for this option.

- **DNS Logs**: Up to eight NSS feeds are supported by ZIA to send the DNS logs to the SIEM.

- **Alert Logs**: ZIA also supports up to eight NSS feeds to monitor the NSS. These alert logs are different from and should not be confused with the alerts described in the previous section. These alerts can indicate whether the connection to the SIEM is lost, or whether the connection to the Zscaler CA is down or the connection to the Nanolog is experiencing any connectivity issues.

- **Tunnel Logs**: If the enterprise has deployed GRE or IPsec tunnels, then up to eight NSS feeds can be configured to send the tunnel logs to the SIEM.

- **SaaS Security Logs**: Up to eight NSS feeds can be configured by the enterprise administrator that can specify many filter conditions, such as a location or a department.

Next, let's find out what kinds of reports can be generated for management review purposes.

Producing reports for management review

From time to time, the enterprise administrator may be called upon by upper management to generate easy-to-understand reports that show whether the Zscaler ZIA solution is working optimally for the enterprise. ZIA Admin Portal makes it easy to create industry-standard and customized reports. Let's explore the built-in reports.

System-defined reports

ZIA Admin Portal offers several types of default, system-defined reports. The two most common choices are the **Executive Report** and the **Industry Peer Comparison Report**. Let's see what is in each type of report.

Executive Report

The executive report contains an overall security view of the enterprise in an HTML format. It shows the value derived from the ZIA service. It contains details such as how many security threats and/or company security policy violations were detected for the enterprise during a certain time frame.

After logging into the ZIA Admin Portal, the administrator can go to **Analytics** > **Reporting** > **Executive Reports**. Once the report loads, the administrator can click on the time frame dropdown on the top left of the page and adjust it as needed. The report thus customized by time frame can then be printed by clicking on **Print** on the top right of the page.

If this report needs to be sent to management contacts on a periodic basis, click on **Schedule** on the top right of the page, just to the left of **Print**. Enter the details for the following options on the pop-up window.

- **Schedule Name**: Choose an appropriate name for this report.
- **Recipients**: Add the email addresses of the management contacts who need to receive this report.
- **Status**: The status of the report can be **Enabled** or **Disabled**.
- **Frequency**: This is a static field that displays the **Monthly** option and cannot be customized. The recipients will receive this report on the first of every month (for the previous month).
- **Time Zone**: The preferred time zone for the enterprise.

Any further changes can be done by clicking on **Edit Scheduled Report** that is shown after a report has already been scheduled. Let's move on to the Industry Peer Comparison Report. Please note that this report will be discontinued and Zscaler recommends using the **Executive Insights Email Report** from the **Analytics** menu instead.

Industry Peer Comparison Report

A second, popular built-in report is the Industry Peer Comparison Report. Often, security tool vendors provide an insight into how an enterprise compares to the industry (such as banking or manufacturing) it is in. This report provides those details and in addition, it also provides a comparison against all organizations using the ZIA service.

To get to this report, the enterprise administrator must log into the ZIA Admin Portal and navigate to **Analytics** > **Reporting** > **Industry Peer Comparison**. If this report looks inaccurate, then the administrator needs to click on the blue **i** icon right beside the **Industry Peer Comparison Report** in the upper-left corner. This shows the **Business Vertical**, **Geographic Region**, and **Business Size** for the enterprise. If this information is not correct, it needs to be corrected by contacting Zscaler.

There are several other types of reports that are available or being added by Zscaler on a regular basis. A list of reports from any given time can be found by logging into the ZIA Admin Portal and navigating to the **Analytics** > **Reporting** section or by consulting the ZIA documentation.

In addition to the built-in reports, the ZIA Admin Portal offers an on-demand query option, called Insights. Let's take a look at the various available Insights options.

Insights

Insights are sort of like on-demand, customizable reports that can be generated by the enterprise administrator. They are especially useful when troubleshooting a specific problem. They can be accessed by navigating to **Analytics** > **Insights**. The most popular Insights option is **Web Insights**.

Once this option is selected, the administrator can adjust the following choices.

- **Timeframe**: The timeframe can be as small as the current day or as large as several days that can be selected using the offered calendar by selecting the **Custom** option. The default option is **Current Day**; other options include **Current Week**, **Current Month**, **Previous Day**, **Previous Week**, **Previous Month,** and **Custom**.

- **Select Chart Type**: The different types of chart visualizations available here are bar, pie, line, or a table format.

- **Units**: The options here are either **Transactions** or **Bytes**.

- **Select Filters**: Multiple filters are available, such as **Department**, **Location**, and more.

Once the administrator makes the necessary adjustments and clicks on **Apply Filters**, the appropriate data is shown on the right-hand side that matches the criteria. This on-demand report can be printed using the **Print** icon on the top right-hand corner of the page. By default, the data shown is for **Overall Traffic** and it can be further narrowed down using available suboptions.

Alternatively, the **Logs** option (in the top-right corner beside the **Insights** option) can be selected to view and download the raw logs that match the same criteria. The timeframe can be narrowed down to as little as the **Last 1 Minute** and as large as the **Previous Month**. There is also a field called **Number of Records Displayed** that can be used to limit the number of records displayed on the screen. Underneath the **Select Filters** option, there are additional options (compared to the Insights options) such as **Client IP**, and **Device Owner**.

Be careful as they may run into several thousand entries. Narrow your criteria as much as you can before proceeding to download the logs in a **Comma-Separated Value (CSV)** format, which can then be used for further offline analysis. The maximum number of records that can be downloaded is 100,000.

Let's now look at how to customize the widgets on the ZIA Dashboard.

Generating custom widgets for the ZIA Dashboard

As soon as the enterprise administrator logs into the ZIA Admin Portal, the **Web Overview** dashboard is displayed by default. This dashboard offers some predefined widgets. These widgets can be edited or deleted, and new custom widgets can be created. Let's take a look at that customization process.

Editing current widgets

The **Dashboard** page is loaded by default when an enterprise administrator logs into the ZIA Admin Portal so, there is no special navigation required after login. Current widgets on the dashboard can be edited by hovering the mouse near the top-right corner of the individual widget to reveal a pencil icon. Click on the *pencil* icon, and you will see two options presented there: **Edit Widget** and **Remove**. Clicking on **Remove** will ask for confirmation before proceeding with the deletion of this widget. Clicking on the **Edit Widget** will open a pop-up menu with the various options. We will discuss those options in the next section.

Adding new widgets

On the top right of the Dashboard page, click on the blue + icon to add a new widget to the dashboard. The options available on the resulting pop-up window are as follows.

The top menu bar specifies the type of data that the widget will display. The default selected option is the **Web**. The other options are **Mobile**, **Firewall**, and **DNS**:

- **Title**: Enter an appropriate title name that will be displayed once the widget is added.

- **Data Type**: There are several options available in the drop-down list – too many to be listed here. Based on the enterprise need, select the most appropriate option. You can also search for an option by typing in the appropriate keyword.

- **Units**: We already saw the choices for this field are either **Transactions** or **Bytes**.

- **Chart Type**: There are four types of chart options, namely, **Bar Chart**, **Pie Chart**, **Line Chart**, **Value Chart**, and **Table Chart**.

- **Filters**: Based on the type of chart, appropriate filter options are displayed and can be optionally selected.

Once all these fields have been entered, click on the blue **OK** button to add this widget to the dashboard. Let's now look at the ZIA troubleshooting guide creation.

Creating a unified ZIA troubleshooting guide

Either through proactive alerting or reactive ticketing, trouble tickets eventually reach the help desk, and they must be worked upon to resolution. When the enterprise adopts a logical and consistent troubleshooting approach, the resolution time for these trouble tickets can be decreased, thus alleviating the pressure on the Zscaler enterprise administrator.

Basic troubleshooting

The basic information that should be gathered by a help desk associate applicable to many common scenarios is as follows.

Access to IP.zscaler.com

If the end user can log into their computer using domain credentials, they should be asked to open a company-approved internet browser and navigate to `ip.zscaler.com`. This can tell us if the end user is accessing the web using the ZIA service or through an alternate path. If the end user is not going through ZIA, this web page will say *"The request received from you did not have an XFF header, so you are quite likely not going through the Zscaler proxy service."*

If the end user is indeed traversing through the ZIA service for their web traffic, the `ip.zscaler.com` page will say *"You are accessing this host via <> hosted at <> in the <cloud name> cloud,"* and provides the following additional details. The end user can then either take a screenshot type in these details into a notepad or an email client:

- The **Zscaler Public Service Edge (PSE)** name
- The egress IP address of the Zscaler proxy
- The virtual IP address of the Zscaler proxy
- The name of the Zscaler proxy
- The egress IP address of the enterprise network
- If authentication is enabled, and if yes, then has the end user authenticated?

Ipconfig or Ifconfig

Depending on the operating system platform of the end user, get them to run the `ipconfig /all` or `ifconfig` command (on Command Prompt in Windows and Terminal in Mac, respectively), which tells us whether the end user has a valid IP address to be able to access the internet via the network they are connected to.

If the end user does not have a valid IP address, check whether their wired or wireless connection says, *"**Media disconnected**"*. If the physical connection is OK, then check whether the end user has a valid IP address as per the enterprise DHCP allocation scheme. If not, check whether the end user can release and renew the IP address lease. In the case of a failure, check the connection to the enterprise DHCP server or engage the appropriate DHCP support team.

If the end user has a valid IP address, check whether they can ping the default gateway. If that does not work, engage the network team or the IT support team as needed. If the ping to the default gateway works, proceed to the next section.

Ping and traceroute (tracert)

In the same Command Prompt or Terminal window, have the end user run a `ping` command and a `traceroute` (`tracert` on Windows) to a well-known URL, such as `www.google.com`, and to the destination IP address to see whether the Layer 3 connectivity and DNS resolution are working. If the ping to `www.google.com` does not work, or the DNS resolution fails, engage the DNS support team.

Advanced troubleshooting

Sometimes, in addition to the steps that we listed in the *Basic troubleshooting* section, Zscaler Support may ask for additional troubleshooting information. Two of the commonly used pieces of information are the HTTP headers and the ZCC App packet capture. Let's gain a better understanding of them now.

HTTP headers

When an end user requests a web page using the HTTP protocol, capturing the HTTP headers from the user request gives us more details, such as the exact URL requested by the end user. The three most popular browsers, namely, Google Chrome, Mozilla Firefox, and Safari, offer additional developer tools that enable this capture. The links to each of these have been provided in the *Further reading* section of this chapter.

ZCC packet capture

The ZCC App offers a packet capture of the end user traffic at a lower level and, hence, captures traffic that may be missed by other packet capture tools such as Wireshark. For the packet capture to work in the ZCC App, the end user must use version 1.3, or later, and this captures the packets at the driver level. The restriction is that only packets that are passing through the filter driver are captured. For ZCC App version 1.5, or later, all packet captures are done at the adapter level. To turn on this feature, and for instructions on how it works, please refer to the link in the *Further reading* section of this chapter.

Now that we are done with the common troubleshooting procedures, let's look at various scenarios that could be referenced in a problem ticket.

End users are unable to access websites

The end user may report that they are unable to access websites. The message they may be getting is that they are not connected to the internet. Follow these troubleshooting steps:

1. Document the website that the end user is trying to access.

2. Have the end user try other websites, such as the ones allowed by the enterprise.

3. Ask the end user whether other users are experiencing the same problem.

4. Check whether the end user's computer has a valid IP address using the commands mentioned earlier in the *Basic troubleshooting* section.

5. If the basic troubleshooting succeeds, have the user retest accessing the website originally reported.

If trouble remains, engage higher-tier support.

End users get a Website Blocked error

In this scenario, the end users accessing certain website categories could be getting a message that says *"Website blocked."* Please follow these troubleshooting steps:

1. Document the website that the end user was trying to access. This can be found a few lines underneath the **Website blocked** error message.

2. Ask the end user whether anyone else is experiencing these error messages for the same website.

3. If possible, check whether this URL is allowed as a per enterprise security policy. If it is indeed prohibited by company policy, inform the end user.

4. If the end user insists, or is in a department that is allowed access to this URL category, then escalate the issue to the Zscaler enterprise administrator.

Next, let's take a look at the captive portal error.

The ZCC App displays a Captive Portal Fail Open Error message

In this scenario, the end user is not at a trusted location and is trying to access the internet from a public Wi-Fi, such as a hotel, airport, or a café. Such public Wi-Fi access points typically ask the end user to acknowledge their **Acceptable Use Policy** (**AUP**) by clicking on **Accept** or **OK**. Follow these steps:

1. Ask the end user whether they got an AUP page or popup that they may have missed.

2. Ask the end user to accept the AUP, and then click on **Retry** inside the ZCC App.

If the issue persists, escalate it to the Zscaler enterprise administrator.

The ZCC App shows a Network Error message

When the ZCC App displays a **Network Error** message, it usually means that an active network interface was not found. Follow these resolution steps:

1. Using the steps from the *Basic troubleshooting* section, check whether the end user has an active network connection, either wired or wireless.

2. Check to make sure the network connection is not in a **Disabled** state. If it is disabled, try to enable it.

3. Once the network interface is enabled, click on **Retry** inside the ZCC App.

4. If the issue persists, attempt to reinstall the driver for the network adapter and click on **Retry** inside the ZCC App.

If trouble persists, engage higher-tier support.

The ZCC App displays an Internal Error message

This error message displayed by the ZCC App usually means that it has encountered an internal socket error. Follow these steps to address this issue:

1. Ask the end user to wait for a few minutes and then click on **Retry** inside the ZCC App.

2. If trouble persists, advise the end user to perform a hard reboot of the computer, which allows the ZCC App to try and perform new socket mapping.

If the issue persists, engage higher-tier support.

The ZCC App exhibits a Connection Error message

A connection error usually means that the ZCC App is unable to reach the Zscaler PSE. Follow these steps to try and resolve this issue:

1. Check the steps in the *Basic troubleshooting* section and see whether the end user can ping and traceroute to `www.google.com`.

2. If the preceding steps succeed, have the end user click on **Retry** inside the ZCC App.

3. If the issue remains, engage higher-tier support to check whether a firewall or a routing rule is blocking access to the Zscaler PSE.

Next, let's examine a local firewall or anti-virus error.

The ZCC App has a Local FW/AV Error message

This error occurs when the ZCC App is being blocked by a local anti-virus application or a firewall. Follow these steps:

1. Ask the end user whether the anti-virus application or signatures were recently updated or check with the IT team if the application updates are centrally administered.

2. Ask the end user if any changes were made to the local firewall.

3. Ask the end users if others are experiencing the same error.

4. If possible, try to disable or turn off the anti-virus and/or the local firewall and click on **Retry** inside the ZCC App.

If the issue persists, engage higher-tier support to see if they can whitelist the ZCC App in the anti-virus and/or the local firewall.

The ZCC App shows a Driver Error message

This error usually means that the ZCC App cannot enable the tunnel interface because of a possible driver installation failure. Attempt these resolution steps:

1. Inside the ZCC App, click on the **More** icon (this is a circle with three horizontal dots inside it) and navigate to the **Troubleshoot** section. Click on **Repair App**.

2. Once the repair process completes, click on **Retry** inside the ZCC App.

If the trouble persists, engage higher-tier support.

User authentication errors

Usually, end users first authenticate using the **Identity Provider** (**IdP**) and **Security Assertion Markup Language** (**SAML**) which enables the exchange of this authentication and authorization information between the IdP and Zscaler. Let's take a look at a family of SAML, **Lightweight Directory Access Protocol** (**LDAP**), and Kerberos authentication errors experienced by the end users.

SAML transit errors

Transit errors are usually temporary and clear on their own. In this scenario, the end user gets one of these error codes – E5503, E5507, E5508, E5611, E5612, E5614, E5619, E5623, E5629, A002, A003, A00C, A00D, A00E, A011, A019, A023, A029, or A02A. Follow these steps:

1. Ask the end user for the duration of this problem and when it worked last.

2. Check whether other users are running into the same issue.

3. Advise the end user to close their browser sessions and attempt authentication after a few minutes.

If the issue persists after several minutes, engage higher-tier support.

SAML account errors

Account errors are encountered by users due to account-related issues such as the account is not activated, disabled, or never provisioned. Check whether the end users report one of these error codes – E5616, E5621, E5624, E5628, or A010. Follow these steps:

1. Check whether the error code seen by the end user matches one in the preceding list.

2. Obtain the username from the end user.

3. Check whether that username exists on the ZIA Admin Portal or the enterprise IdP. If that username is not found, engage the IdP team or the Zscaler administrator.

4. If the username exists, make sure the user's account is not disabled. If it is disabled, engage the appropriate team to see if, and how, it can be re-enabled.

5. Check whether the auto-provisioning option feature is enabled on the ZIA Admin Portal.

If trouble persists, engage higher-tier support.

SAML login format errors

Login format errors occur when end users do not use the proper format for their username during authentication. In these scenarios, they get an error code of A021. Follow these resolution steps:

1. Ask the end user what username they are using to authenticate.

2. If the end user does not use an email address format for their username, please provide the end user with the correct username format and retry authentication.

If trouble persists, engage higher-tier support.

LDAP password errors

When the enterprise is using LDAP as the configured authentication method and end users experience authentication errors due to a wrong password, they usually get an error code of 101. Follow these troubleshooting steps:

1. Ask the end user whether they recently changed or reset their password.

2. Have the end user retry authentication using the most recent password.

3. If necessary, reset the end user's password and have them retry authentication.

If trouble persists, engage higher-tier support.

LDAP account errors

If the enterprise is using LDAP and the end user account is not activated or found on the LDAP server, then they will experience account-related errors. They might get one of these error codes – 103, 106, or 113. Follow these resolution steps:

1. Check whether the end user account exists on the LDAP server. If not, engage the LDAP support team.

2. If the end user account is found on the LDAP server but is in a disabled state, try getting the account enabled and have the user retry authentication.

If trouble persists, engage higher-tier support.

LDAP transit errors

LDAP transit errors usually clear on their own. In these situations, the end users get one of these error codes – 102, 107, 111, 114, or 115. Use these troubleshooting steps:

1. Ask the end user when the last time the authentication was working properly.

2. Ask the end user whether others are experiencing the same problem.

3. Advise the end user to close all browser windows and retry authentication after a few minutes.

If trouble persists, engage higher-tier support.

Kerberos account errors

When the enterprise is using Kerberos as the authentication method and the user's account is not activated or found on the server, they can experience account authentication errors with one of these error codes – 441000 or 461000. Use these troubleshooting steps:

1. Obtain the username from the end user that is being used for authentication.

2. Check whether the end user account exists on the server. If the account does not exist, engage the appropriate support team.

3. If the account exists on the server, but it is in a disabled state, engage the appropriate support team to get it enabled.

4. Have the end user retry authentication.

5. If trouble persists, engage higher-tier support.

So far, we have looked at SAML, LDAP, and Kerberos authentication errors. Let's now examine some more trouble scenarios that might be seen by the help desk.

Users are unable to upload or download files

In this scenario, end users report the inability to download or upload files to certain websites such as Google Drive or Dropbox. This could happen either due to a ZIA-enforced policy or a browser compatibility issue. Follow these troubleshooting steps:

1. Ask the end user when the last time was they were able to perform this action and if it ever worked for them. If it never worked, this could be by design.

2. If it worked in the past and not working now, obtain the error message or Zscaler block message displayed on the user's screen.

3. Check what action is being blocked – upload, download, or both.

4. Check whether other users in the same workgroup, department, or location can perform the same action(s) successfully.

5. Ask the end user to try the same action using a different browser.

6. Check whether this block action is applicable to a specific file type and/or size.

Engage the next tier support or Zscaler administrator as needed.

Slow website response

In this scenario, end users report slow loading websites or erratic video streaming. This could be caused due to browser compatibility issues or due to a bandwidth enforcement policy of the enterprise. Follow these steps to troubleshoot these issues:

1. Ask the end user whether the slowness is with certain website categories or a particular website.

2. Check whether multiple end users are experiencing that same slowness.

3. Ask the end user to try accessing the same website using a different browser and a different computer.

4. Perform traceroute to a website without a slow response and compare it to a website that experiences slowness and compare the results.

If trouble persists, engage higher-tier support or a Zscaler administrator so that the link bandwidth can be checked for contention or a Zscaler bandwidth enforcement policy.

In addition to the issues seen by the end users, it is worth mentioning a few issues that could be seen by ZIA administrators.

URL formatting

When the ZIA administrator is whitelisting certain domains, it is important to understand the behavior of non-Zscaler proxy solutions such as Bluecoat or other similar products. In those products, if a domain (such as `example.com`) is whitelisted, the product automatically and implicitly whitelists all the subdomains (such as `download.example.com` and `mail.example.com`).

However, in ZIA, if you need to whitelist multiple subdomains, you need to specify a dot or a period (`.`) to the left of the URL, and it will allow up to five subdomains deep. Failure to understand the behavior could result in the accidental blocking of the intended subdomains.

Application SSL inspection

The ZIA administrators need to be aware of the behavior of the URL and firewall filtering policies when SSL inspection is turned off for a specific location or a sublocation. In such cases, the firewall and URL filtering policies may fail intermittently.

Application authentication

Many legacy enterprise applications do not support SAML or interactive authentication. As such, it is not recommended that you use service accounts as the username as the application authentication may fail inside the proxy. In these cases, the solution is usually to bypass authentication and create a separate location or sublocation for such sources.

Summary

In this chapter, we looked at the various built-in tools provided by the ZIA Admin Portal, by default, and customizable tools to suit the needs of the enterprise operations staff. They come in various forms such as Insights, Logs, Dashboards, Reports, and more.

We also looked at some of the most frequently seen troubles with the ZIA end users and how a streamlined troubleshooting approach can help the enterprise resolve them effectively.

In the next chapter, we will start addressing the **Zero Trust Network Architecture (ZTNA)** using **Zscaler Private Access (ZPA)**.

Questions

As we conclude, here is a list of questions for you to test your knowledge regarding this chapter's material. You will find the answers in the *Assessments* section of the Appendix:

1. ZIA Admin Portal dashboards cannot be customized by the enterprise administrators.

   a. True

   b. False

2. Using the **Logs** option, the enterprise administrator can download the raw logs for the selected transaction criteria.

   a. True

   b. False

3. User authentication issues occur due to the following reasons:

   a. End user errors

   b. Misconfiguration errors

   c. Provisioning errors

   d. None of the above

   e. All the above

Further reading

- Capturing HTTP Headers on Google Chrome: `https://help.zscaler.com/zia/capturing-http-headers-google-chrome`

- Capturing HTTP Headers on Mozilla Firefox: `https://help.zscaler.com/zia/capturing-http-headers-mozilla-firefox`

- Capturing HTTP Headers on Safari: `https://help.zscaler.com/zia/capturing-http-headers-safari`

- Enabling Packet Capture for ZCC App: `https://help.zscaler.com/z-app/enabling-packet-capture-zscaler-app`

- URL Format Guidelines: `https://help.zscaler.com/zia/url-format-guidelines`

Section 2: Zero-Trust Network Access (ZTNA) for the Modern Enterprise

This part will detail how ZPA can be used to deploy a zero-trust network access solution for the modern enterprise's cloud-based and legacy applications.

This section comprises the following chapters:

7
Introducing ZTNA with Zscaler Private Access (ZPA)

In the very first chapter, we discussed two main issues faced by modern enterprises: the first being the need for a secure web access experience, and the second being the need to connect enterprise applications to their end users in a very flexible, secure way.

The first issue was addressed by the **Zscaler Internet Access (ZIA)** offering from Zscaler. This chapter discusses in detail the second offering from Zscaler, which is **Zscaler Private Access (ZPA)**. We will look at the need for a **Zero Trust Network Access (ZTNA)** solution that is satisfied by ZPA. We will also look at how ZPA is architected, along with its related components.

You will learn about the need for ZTNA in today's enterprise environment, which drastically rethinks the security access model. You will also understand the building blocks of the ZPA architecture, and the role played by each component. We will also explore some clientless ZPA solutions where it is not possible to use the **Zscaler Client Connector (ZCC)** app.

In this chapter, we are going to cover the following topics:

- What is ZTNA and how does ZPA fit in to this?
- Delving into the ZPA architecture
- Exploring clientless ZPA solutions

What is ZTNA and how does ZPA fit in to this?

ZTNA is defined by Gartner (`https://www.gartner.com/en/information-technology/glossary/zero-trust-network-access-ztna-`) as *"a product or service that creates an identity- and context-based, logical access boundary around an application or set of applications. The applications are hidden from discovery, and access is restricted via a trust broker to a set of named entities. The broker verifies the identity, context and policy adherence of the specified participants before allowing access and prohibits lateral movement elsewhere in the network. This removes application assets from public visibility and significantly reduces the surface area for attack."*

This definition by Gartner is certainly a mouthful and is very generic in nature. Let's adapt this definition to the one given by Zscaler, as it applies to its ZPA solution. Zscaler defines this as follows (see `https://www.zscaler.com/resources/security-terms-glossary/what-is-zero-trust-network-access` for more details):

> *"Zero trust network access (ZTNA), also known as the software-defined perimeter (SDP), is a set of technologies that operates on an adaptive trust model, where trust is never implicit, and access is granted on a "need-to-know," least-privileged basis defined by granular policies. ZTNA gives users seamless and secure connectivity to private applications without ever placing them on the network or exposing apps to the internet."*

The ZTNA definition by Zscaler is a little bit easier to understand for anyone who is familiar with the security terms used in that definition. A traditional firewall, **Virtual Private Network** (**VPN**), or a proxy generally creates network segments—such as trusted segments and untrusted segments, but the ZTNA framework takes a totally different approach.

ZTNA core principles

Zscaler breaks this down into the following four core principles (taken from the same **URL** as previously):

1. ZTNA completely isolates the act of providing application access from network access. This isolation reduces risks to the network, such as infection by compromised devices, and only grants application access to authorized users.

    As mentioned previously, application access no longer means access to the underlying network that the application resides on. This means that lateral movement by a compromised device is no longer possible because all the compromised device sees is the application and not the network.

2. ZTNA makes outbound-only connections, ensuring both network and application infrastructure is made invisible to unauthorized users. **Internet Protocol (IP)** addresses are never exposed to the internet, creating a *darknet* that makes the network impossible to find.

    When end users request access to an application, both the end users and the applications have *fake* IP addresses that are never exposed to the internet. This means that whoever is scanning for the IP addresses on the public internet can never find these IP addresses, so there is nothing to attack.

3. ZTNA's native app segmentation ensures that once users are authorized, application access is granted on a one-to-one basis. Authorized users have access only to specific applications, rather than full access to the network.

    Although this principle looks very similar to the first one mentioned previously, this principle combines the least-privilege-access concept so that end users have just enough access to perform their work and cannot expand their access scope to beyond what is already provisioned.

4. ZTNA takes a user-to-application approach to security, rather than a network-centric approach. The network becomes deemphasized and the internet becomes the new corporate network, leveraging end-to-end encrypted **Transport Layer Security (TLS)** micro-tunnels instead of **Multiprotocol Label Switching (MPLS)**.

    As explained earlier in the first principle, an end user is granted access just to an application, not the underlying network. This connection between the end user and the application occurs via encrypted TLS micro-tunnels that are established through the public internet; so, each end user-to-application session is an encrypted TLS micro-tunnel of its own.

Now that we have defined what ZTNA is and its four core principles, let's explore the need for ZTNA in today's enterprise security solutions.

Why is ZTNA needed?

Let's explore why enterprises today are looking toward adopting ZTNA. Here are some reasons for this:

- **Alternative to traditional VPNs**—VPNs have become slow and difficult to administer and manage, and offer limited security. Gartner predicts the following: *"By 2023, 60% of the enterprises will phase out most of their remote access VPNs in favor of ZTNA."*

- **Secure access to multiple clouds**—As many enterprises are adopting the cloud and migrating to multiple cloud providers for redundancy and portability, ZTNA is becoming increasingly popular.

- **Least-privilege access**—Third-party vendors and users often receive over-privileged access, creating security risks and gaps for an enterprise. ZTNA significantly reduces this risk by making sure that external users only gain access to authorized applications instead of gaining access to the network hosting those applications.

- **Faster integration**—When an enterprise acquires one or more businesses through mergers and acquisitions, network integration between the enterprise and the new companies can take years, with overlapping IP addresses. ZTNA simplifies and reduces the time and management needed for this effort, and yields quick business value.

Now that we have understood what ZTNA means and the need for it, let's examine how the ZPA product offering satisfies these ZTNA conditions.

ZPA security principles

The ZPA product offering was created with the four ZTNA core principles explained earlier. These are outlined again here:

Principle 1: Provide end users with access to the applications, not to the underlying network. ZPA only provides end users access to a specific application based on policy rules, without placing the end user onto the private network.

Access to applications should not be dependent on network access, such as **access control lists** (**ACLs**) based on IP addresses. If network access is not provided to end users, there is no risk of lateral movement (in case of any compromised devices) throughout the network, which may consist of hundreds of other resources. This minimizes the attack surface, providing a better level of security.

Principle 2: Applications are never exposed to the end users. ZPA does not advertise application availability in general to everyone. Applications are only visible to authenticated and legitimate end users.

By restricting the discovery of internal enterprise applications only to authenticated users, unauthenticated users who form a majority are unable to view those applications, thereby preventing any possible attack or exploitation. Not using inbound connections or public IP addresses creates an enterprise darknet, whereby applications are not exposed to the internet.

Principle 3: Don't segment the network—segment your applications. ZPA segments applications based on who the users are, what their access entitlement is, and the security-posture status of their end device.

When mapping is performed from a specific end user to a specific application using a per-session TLS micro-tunnel (thereby moving to a user-to-application model from a network-to-network model), this removes any lateral movement on that secure connection.

Principle 4: Use the internet for remote access, and not a VPN. ZPA leverages the internet by using secure TLS micro-tunnels to access private enterprise applications, without opening network-level access. ZPA also allows for optional double encryption of data transfer between the end-user device and the application, for complete security and privacy.

There is no need for a VPN in a ZPA solution. Both the end user and the application endpoints establish dynamic outbound TLS tunnels to the Zscaler cloud, and the Zscaler cloud will then broker the secure end-to-end connection between the end user and the application. Data is already encrypted end to end, and—optionally—enterprises can use their own client and server certificates for double encryption.

Now that we have looked at the four core security principles of ZPA, let's explore the architecture of ZPA.

Delving into the ZPA architecture

ZPA only supports communications in the client-to-server direction. Any other models (such as server-to-client, client-to-client, and so on) are not supported by ZPA. Important components of the ZPA architecture are the ZPA **Central Authority** (**CA**); ZPA **Public Service Edge** (**PSE**); ZCC application; App Connectors; **ZPA Tunnels** (**Z-Tunnels**); **Microtunnels** (**M-Tunnels**); the logging and analytics cluster; and the **Log Streaming Service** (**LSS**).

Let's look at each of the components in detail, beginning with the ZPA CA.

ZPA CA

We already looked at the CA when we learned about the ZIA architecture in *Chapter 2, Understanding the Zscaler Modular Architecture*. Similarities between both the ZIA and ZPA are that both are multi-tenant, redundant, globally distributed policy engines. However, the main difference is that the central purpose of the ZPA CA is to enable connection requests, in addition to enforcing provisioning policies.

Just as with the ZIA CA, the ZPA CA contains the configuration and policy that has been put in place by the enterprise administrator, based on the business needs. The CA also offers detailed visibility into end-user activity and application access. Where the ZPA CA differs is that it supports the App Connectors with their configurations.

To support end-user and administrator authentication, one or more **Security Assertion Markup Language** (**SAML**) **identity providers** (**IdPs**) must be configured on the ZPA CA, and—if necessary—customer-signed enrollment certificates may also be loaded onto the ZPA CA. The ZPA CA also lists applications that were discovered and can be configured to apply policies for specific applications defined by hostname or IP address and port range.

Manual or dynamic discovery of the servers hosting the applications is supported, although the dynamic option is recommended to reduce manual overhead. The ZPA CA also monitors the reachability and health of these applications so that end users are always connected to the best possible instance of the application. Granular policies can also be configured to control exactly which applications can be accessed by which end users.

Let's explore the next component of the ZPA architecture: the ZPA PSEs.

ZPA PSEs

ZPA PSEs are globally available nodes that broker the connection between end users and connectors, for specific application access. Again, note that ZPA PSEs are separate from ZIA PSEs because they serve completely different functions. Also, ZPA PSEs are a combination of Zscaler in-house data centers, **Amazon Web Services** (**AWS**), and Azure resources, to provide global optimum coverage.

End users approach the ZPA PSEs, looking to connect to the applications they are authorized for. The destination applications are already discovered by the App Connectors, and the App Connectors have already approached the ZPA PSEs, effectively announcing that they are ready and available for the end users. The ZPA PSEs then broker the end-to-end connections between these end users and the application connectors.

In addition to brokering these connections, the ZPA PSEs also provide authentication, end-user policy enforcement, and secure data forwarding through single encrypted and—optionally—double-encrypted TLS tunnels.

Let's now look at the next component of the ZPA infrastructure: the ZCC application.

ZCC application

We already spent a lot of time and effort to understanding the flexibility of this application for ZIA access. The ZCC application forms the central component for end users to access applications using ZPA. When the end-user device is connecting to ZPA applications, the browser or agent on the end-user device thinks it is talking directly to the destination application using a synthetic, *fake* IP address assigned to the application by the ZCC app.

When Z tunnels are established to the nearest ZPA PSE and microtunnels are eventually established between the end-user device and the destination application, the ZCC app serves as the origin for both tunnels. Posture profiles can be defined and associated with the end-user device so that access to the application is only allowed when the end-user host device complies with the specified posture requirements.

Upon successful enrollment with the ZPA PSE, the ZCC application is issued an identity certificate that is signed by the subordinate CA. The keys and certificates are securely stored within the ZCC app itself, and they are renewed every time the end user re-enrolls into the app. In the case of any indication of compromise, the enterprise administrator can disable ZPA access for that end device, and the certificates are automatically revoked.

When the ZCC app enrolls with the ZPA PSE, a hardware fingerprint is also captured with unique data from the end device, to prevent the possibility of cloning the ZCC app and its certificates by unauthorized parties. This unique data consists of the **Central Processing Unit** (**CPU**) ID, the unique battery ID, and so on (among other items). This fingerprint is validated every time the app checks in with the ZPA PSE. The ZPA CA then notifies the ZCC app of the available private enterprise applications and the applications that require double encryption.

ZPA web application access can also be achieved using a standard browser, with a **Browser Access** (**BA**) configuration option. The ZCC app allows additional features such as an end-user-device posture check and validation using a trusted network configuration and provides multi-protocol access to destination private applications. The BA option is only limited to web applications and will be covered later in this chapter.

When the end user requests access to an application by using the IP address or the **Fully Qualified Domain Name** (**FQDN**), the ZCC app will handle that request and establish a local synthetic IP address for the outbound connection. These IP addresses are from the **Request for Comments** (**RFC**) 6598 carrier grade range of `100.64.0.0/10`, with a ZCC app using a modified subnet mask of `/16`. The client simply sends all requests to this allocated synthetic IP address.

This request is then sent by the ZCC app to the ZPA PSE, to validate the requested application and the TCP/UDP port (where **UDP** stands for **User Data Protocol**) being requested and provide SAML assertion for the user. The ZPA PSE then queries the ZPA CA to verify if this access is allowed per the defined policies of the enterprise, and checks if the application is still available before selecting the best possible connector.

Let's now explore the next component of the ZPA architecture: the App Connectors.

App Connectors

App Connectors are usually the **Remote Package Manager** (**RPM**) or **Virtual Machines** (**VMs**) installed on the destination network (typically on the same subnet as the applications) hosting the applications. On bootup, these VMs establish an outbound encrypted TLS connection to the nearest and healthy ZPA PSE. This connection is the control plane through which the ZPA PSE controls the configuration of the connector and announces the availability of the discovered application to the ZPA CA when needed.

Connectors do not need—or even support—any inbound connections. They serve as the origin point for the application Z tunnels to the Zscaler cloud and as the destination for the end-to-end microtunnels described previously.

Connectors must be able to resolve the applications via the **Domain Name System** (**DNS**) so that they can both connect to those applications and then can be made available over ZPA. In turn, the applications believe that they are simply talking to the connector and are unaware of the tunneling that will be used to connect with the end user's device.

A suitable ZPA subsidiary CA usually signs a provisioning key, which is then needed to enroll a connector. After this enrollment, the connector receives its configuration and certificates from the ZPA CA. These connectors are then controlled from the ZPA CA and can be updated or restarted as needed. Also, a device fingerprint is captured with unique data from the host device, to prevent any cloning of the connector and its certificates by unauthorized parties.

Each connector has throughput support of up to 500 **megabits per second** (**Mbps**). If additional capacity is needed by the enterprise, the connectors need to be scaled horizontally by adding more connectors. No special load balancers or clusters are needed. The ZPA intelligently and automatically distributes the originating user sessions across the available connectors to ensure the best user experience.

To avoid a complete outage in case a particular connector is being updated or restarted, it is recommended that connectors be deployed in pairs for redundancy and high availability. The connectors usually undergo weekly software updates.

Let's move on to the next component of the ZPA infrastructure: the Z tunnels.

Z tunnels

Z tunnels are fully encrypted TLS tunnels on port 443 that are both established outbound initially between the end-user ZCC App and the ZPA PSE (initiated by the end-user ZCC App) and then between the app connectors and the ZPA PSE (initiated by the connectors). These tunnels are double-pinned and mutually validated, providing immunity against **Man-in-The-Middle** (**MiTM**) attacks.

Z tunnels from the ZCC app to the ZPA PSE use **Multi-Factor Authentication** (**MFA**). They use SAML assertion, a user identity certificate, and a hardware fingerprint for this authentication. Similarly, Z tunnels from the connectors to the ZPA PSE use an identity certificate and a hardware fingerprint for validation. These tunnels are established using TLS version 1.2 and use the strongest encryption cipher that is mutually supported by the ZCC app, connector hosts, and the ZPA PSEs.

Let's examine the next component of the ZPA architecture: the microtunnels.

Microtunnels

Microtunnels are end-to-end data-byte stream connections, established between the end user and the destination application, originating from the end-user device, traversing through the ZPA PSE and the App Connector, and finally terminating on the destination application. Microtunnels are created on a per-user to per-application basis and cannot be shared with another user or application.

In an MPLS network, the carriers create unique IDs for the various interfaces on devices. This allows the devices to use those ID numbers when exchanging data packets between themselves. When a data packet eventually leaves the MPLS network, the last device on the path removes all the MPLS IDs before sending the data packet to its destination, which is not MPLS-aware.

Similarly, unique IDs are dynamically allocated by the ZCC app and the App Connectors, and those IDs are used to establish the session. The ZPA PSE is aware of these IDs and simply swaps the IDs, switching the traffic as needed between the ZCC app and the App Connectors. The end-user device and the applications are both unaware of this ID mechanism.

There is an additional option of double encryption. A second tunnel can be established within the microtunnel, using the strongest encryption cipher mutually supported by the ZCC app and the App Connectors. If customer keys are used for encryption, then even the ZPA PSE cannot intercept—or even read—the data within that second tunnel. This results in the data being double-encrypted between the ZCC app and the App Connector, providing an enterprise a greater level of security if there is such a business need.

Let's look at the next component of the ZPA infrastructure: the logging and analytics cluster.

Logging and analytics cluster

As with the logging options that we saw with ZIA, ZPA also offers real-time visibility into logs generated by various operations, such as the following logs:

- **Primary tunnel logs**—These logs capture authentication events between the App Connectors and the ZCC app on the end-user devices.

- **Microtunnel logs**—These logs capture user-data transactions.

As with ZIA logs, ZPA logs do not contain any end-user **personally identifiable information** (**PII**). The logs are not stored on any persistent hardware media until they arrive at the log cluster. The end-user information can also be obfuscated based on the needs of the enterprise.

The data that is stored at rest at the log cluster has abstracted IDs and does not make sense on its own. This information can be decoded only when combined with additional information stored somewhere else.

Let's look at the final component of the ZPA architecture: the **Log Streaming Service** (**LSS**).

LSS

LSS is very similar to the **Nanolog Streaming Service** (**NSS**) that we discussed in ZIA. LSS allows an enterprise to stream data—such as user activity, user and connector authentication, and BA logs—to the enterprise **Security Information and Event Management** (**SIEM**) for analysis and correlation, and to integrate it with a chosen ticketing platform. LSS needs a connector adjacent to the SIEM. LSS streams the logs through a ZPA PSE, which forwards it to a log receiver (SIEM) through a suitable connector.

In summary, we have seen that both the ZIA and ZPA PSEs are co-located within the same Zscaler data centers but serve very different purposes. We also saw the role and importance of each of the ZPA components and how they work together to provide end users secure access to enterprise private applications without ever exposing either to the internet.

Exploring clientless ZPA solutions

We looked at the ZPA architecture featuring the ZCC app. In certain environments, situations, and platforms, the ZCC app cannot be supported or installed. Let's look at two such clientless ZPA solutions.

Understanding the Zscaler Cloud Connector ZPA solution

Zscaler Cloud Connector aligns with the zero-trust access philosophy. It is a cloud-native service that allows for fast, secure connectivity between apps, and between an app and the internet.

Cloud connector

The **cloud connector** itself is a software instance that is in front of a VPC in AWS or a **virtual network** (**VNET**) in Microsoft Azure. Just as with the App Connector establishing outbound **Datagram Transport Layer Security** (**DTLS**) connections to the ZPA cloud, these cloud connectors establish outbound DTLS connections to a connection broker in the **Zero Trust Exchange** (**ZTE**).

ZTE

The **ZTE** is a large security cloud with a global footprint of more than 150 Zscaler data centers. This is like the Zscaler ZIA and ZPA cloud, and delivers a single control plane for secure traffic forwarding between end users and the internet, and between applications themselves.

Delving into the BA ZPA solution

A second—and more popular—clientless option is the **BA (Browser Access)** clientless ZPA solution.

This solution is designed for ZPA applications that can be accessed by end users using a standard web browser. Both end-user authentication and eventual application access is all done using the web browser, without the need to install the ZCC app on end-user devices. There is also no need for a browser extension or a plugin, and not even a Java client on the end-user device.

This solution is designed for third-party vendors and users such as contractors who are unable to install the ZCC app on their end devices. This solution can also be used for platforms such as Google Chromebooks or Linux devices, as the ZCC app is not yet supported for those platforms.

BA components

The BA architecture consists of four components—namely, BA Exporter, BA Certificate, BA DNS CNAME Record, and the BA Crypto Store. Let's understand their role in detail.

BA Exporter

The **BA Exporter** is a secure web proxy that is located just before the ZPA PSE, and its primary purpose is to listen for incoming BA application requests. Upon receiving such a request, the BA Exporter responds with the needed BA Certificate and establishes a Z-tunnel to the local ZPA PSE upon successful end-user authentication.

BA Certificate

The **BA Certificate** is a web server certificate for one or multiple BA applications. It can also be a wildcard certificate.

BA DNS CNAME Record

The **BA DNS CNAME Record** is a CNAME alias for a BA application that is usually resolved to the best BA Exporter.

BA Crypto Store

The **BA crypto store** is a key store for the BA that contains the BA Certificate private keys and is in the Amazon **Key Management Service** (**KMS**).

BA workflow

Once the applications have been identified for migration to ZPA, a BA certificate for each such application must be provisioned in the ZPA cloud. To facilitate the DNS resolution of the application FQDN to the ZPA infrastructure, a CNAME record must be defined for each application on the DNS, typically on the external public DNS servers.

Once these steps have been completed, the end user requests access to the ZPA application by entering the FQDN as a URL in the address bar of the web browser. The DNS server will find a matching A record and will resolve it to an IP address. The user's device then receives the IP address for the best possible BA Exporter that the end user can use to then access the requested application.

The user's browser session connects to this BA Exporter resolved by DNS and then receives the BA certificate (already uploaded to the ZPA cloud) for that application. If the user device already has the proper root CA certificate, this connection to the BA Exporter is then trusted and allowed to proceed.

A TLS connection is then established between the end-user browser and the BA Exporter, and end-user authentication using SAML and the configured IdP is initiated. The user proceeds to authenticate at this point using their credentials, and receives a SAML assertion upon successful authentication. This is no different from the ZPA authentication process using the ZCC app.

The access policy is then checked and, if allowed, the user is granted access to the requested application. The ZPA cloud then provides a combination of the best possible ZPA PSE and the best possible App Connector to access the application. The second leg of the connection is established between the BA Exporter and the App Connector, all the way to the ZPA application.

The final step is the creation of the end-to-end microtunnel, providing an encrypted data path from the user's browser to the application. **HyperText Transfer Protocol Secure (HTTPS)** is used to encrypt the session between the browser and the BA Exporter, and Z tunnels are used for the rest of the connection path. This microtunnel is created in a unique per-user and per-application combination. It cannot be used by another user or for accessing another application.

BA certificate creation

Let's walk through the process of creating a BA certificate (typically a web certificate from a private CA owned by the enterprise). The first step is to log in to the ZPA Admin Portal. Click on **Administration** -> **Certificate Management** -> **Browser Access Certificates**. Then, click on the **Browser Access Certificates** tab at the top of the page. Click on the **Create CSR** option in the top right-hand corner of the page to start a new **certificate signing request (CSR)**. On the **Create CSR** popup that appears, fill in the following options:

- **Name**—A name for this CSR.
- **Description**—A free-flowing text field that can be used to describe the purpose of this CSR.
- **Subject**—This field needs to conform to the X.520 syntax for the **Relative Distinguished Name (RDN)**, with the following notation: C (Country), ST (State), O (Organization), CN (Common Name), and so on.
- **Subject Alternate Name**—If necessary, fill in this field; it should match the CN field used in the previous **Subject** field.

Click **Create** to create this CSR. The next step is to download the CSR file and get it signed by the enterprise private CA. Follow these steps:

Click on the *pencil* icon against the CSR just created, to edit the CSR. You can either click on the **Download CSR** option or select the text for the CSR into a text editor.

The next step is to log in to the enterprise root CA and select the option to request a certificate. You should choose the right format when submitting this certificate request (such as Base 64-encoded). Choose the CSR contents saved previously and generate a web-server type of certificate. Download the generated certificate using Base-64 encoding and save it to the computer from where you are logged in to the ZPA Admin Portal.

Back on the ZPA admin portal, you should have the **Create CSR** popup window still open, or you can navigate to it as instructed previously. Under the **Certificate** section, click the **Select File** button and upload the certificate file that was saved to the computer just now. This completes the process of creating a BA certificate.

Creating a BA CNAME record

After generating the BA certificate, let's walk through the process to create the CNAME record for BA. The administrator needs to log in to the ZPA admin portal and click on **Administration** -> **Application Management** -> **Application Segments**. Under the **Application Segments** tab on the page, select the chosen application segment (by clicking on the blue *pencil* icon against it) to enable that application for BA.

On the resulting **Edit Application Segment** pop-up window, navigate to the **Applications** section and click the checkbox that says **Browser Access**. You will see additional options underneath the selection. Select the BA certificate name that was created in the previous section, and then select **HTTPS** under **Protocol** so that the BA Exporter can use HTTPS when connecting to the destination ZPA private application. You may change the port number (optionally) if it deviates from the default, and check the box that says **Use Untrusted Certificates** if the application server certificate cannot be validated by the BA Exporter. Click **Save** to apply your changes.

As soon as you enable BA for an application, it appears under the **Browser Access** tab at the top of the page. Navigate to that tab and click on the chevron (>) sign to the left of your application name to view the details. Copy the CNAME, and then add it to your DNS server. This completes the CNAME configuration for BA.

Testing BA

Once we are done with the BA certificate creation and the CNAME update process, it is time to test BA from the end user's computer. Have the end user log in to their computer and make sure the ZCC app is not running. If it is running, have the user log out of the ZCC app and close the app completely.

Have the end user launch a browser window and navigate to the web application that we configured for BA earlier. The user should be prompted to authenticate using the IdP configured by the ZPA administrator. Once the user authenticates using their credentials, the browser should immediately load the web application using BA. To verify the certificate presented by the ZPA cloud, have the user click on the *lock* icon to the left of the URL. Click on **View Details**, and it should show that this certificate has been verified by the enterprise private CA and can be fully trusted.

Summary

In this chapter, we learned about the necessity for ZTNA and its core principles. We also saw how ZPA aligns with the core principles of ZTNA and provides end-user access to enterprise private applications.

In the next chapter, we will review the various types of system-provided dashboards offered by the ZPA Admin Portal. We will also explore the most common basic configuration options that need to be performed by any ZPA enterprise administrator.

Questions

As we conclude, here is a list of questions for you to test your knowledge regarding this chapter's material. You will find the answers in the *Assessments* section of the *Appendix*:

1. ZPA aligns with the core principles of ZTNA.

    a. Yes

    b. No

2. Enterprise end users access applications over ZPA using a publicly routed IP address.

    a. Yes

    b. No

3. ZPA can only be supported by the ZCC agent.

    a. True

    b. False

4. ZIA and ZPA use the same components and infrastructure components.

    a. True

    b. False

Further reading

- ZCC: https://www.zscaler.com/products/zscaler-cloud-connector
- ZPA BA: https://help.zscaler.com/zpa/about-browser-access

8
Exploring the ZPA Admin Portal and Basic Configuration

In this chapter, we will navigate through the **Zscaler Private Access** (**ZPA**) Admin Portal and configure the ZPA log servers. We will also discuss the steps involved in configuring authentication and the **Zscaler Client Connector** (**ZCC**) app for ZPA.

These topics lay the foundation needed for an administrator to deploy a ZPA solution within an enterprise. In this chapter, we are going to cover the following topics:

- Navigating around the ZPA Admin Portal
- Configuring the ZPA log servers for activity insights
- Integrating with **Azure Active Directory** (**Azure AD**) and Okta for **single sign-on** (**SSO**)
- Configuring the ZCC app for ZPA

Navigating around the ZPA Admin Portal

Just as the **Zscaler Internet Access** (**ZIA**) Admin Portal was used to configure the feature and policy settings for end-user secure web access, the ZPA Admin Portal can be used to configure the various ZPA components and the user access policy.

Unlike the ZIA Admin Portal, which was unique to each cloud name, the ZPA Admin Portal is just one **Uniform Resource Locator** (**URL**), and that is `https://admin.private.zscaler.com`. Once an enterprise signs a contract with Zscaler, the login credentials for the default administrator are provided by Zscaler to the enterprise.

The enterprise administrator then logs in to the ZPA Admin Portal using these credentials, accepts the **end-user license agreement** (**EULA**), and immediately changes the Zscaler-provided password to a very strong password, as mandated by the enterprise security policy.

Once logged in to the ZPA Admin Portal, the primary navigation categories are as follows:

- **Dashboard**—This category consists of the various real-time dashboards that provide an overview of the enterprise applications and users, and overall system health.

- **Diagnostics**—This category provides a page where an enterprise administrator can review and search ZPA logs.

- **Live Logs**—This category is like the ZIA logs, where real-time end-user and connector activity can be seen.

- **Administration**—This category provides multiple options and their related sub-options that allow an enterprise administrator to configure the various ZPA components.

- **Search**—This category offers search options to find tabs, field names, tool tips, and so on.

- **ZCC App**—This category is focused exclusively on the ZCC App portal to configure the environment for the ZCC app users of an enterprise.

In the bottom left-hand corner of the page are housekeeping options that allow the viewing of the currently logged-in user's username, along with a link to access the ZPA **Help Portal** and a **Logout** option. Let's start by exploring the dashboards in detail.

ZPA dashboards

ZPA offers three built-in dashboards—namely, the **Applications** dashboard, the **Users** dashboard, and the **Health** dashboard. Recall that the primary purpose of ZPA is to connect destination applications to their originating end users. So, the first dashboard is the **Applications** dashboard. Let's explore it in more detail.

Applications dashboard

This dashboard is displayed by default (on the landing page) when an administrator first logs in to the ZPA Admin Portal. This dashboard provides information about the enterprise applications either defined by the administrator or automatically discovered by ZPA. By default, it shows the number of applications accessed, the number of discovered applications, the number of access-policy blocks, and the number of successful transactions.

An administrator can also view a list of applications accessed by end users; the top applications by bandwidth; the top errors encountered; the top policy blocks; the top applications by end-user usage; as well as a list of discovered applications. These widgets can be customized by an administrator as per the enterprise needs. By default, the timeframe in the top-middle of the page is set to **24 hours**, but this can be changed to select one of the following options: **1 hour**, **48 hours**, **3 days**, **5 days**, **7 days**, **10 days**, **14 days**, and **Custom Range**.

Hovering the mouse over the name of an application provides a popup with more detailed options, such as the last time the application was accessed, the username of the user who accessed it, the user's **Internet Protocol** (**IP**) address, and the user's location. Clicking on the application name opens that application in the **Diagnostics** section, to perform further filtering.

Moving the mouse over an application name under the **Top Applications by Bandwidth** section displays a popup with additional information, such as the bandwidth used during the timeframe, and the value as a percentage of the total traffic.

Mousing over an error in the **Top Errors** section and the **Applications** tab provides a popup with the total number of transactions and the percentage of these errors compared to the total errors. Further hyperlinks are provided to analyze this data by connectors and the connection status code, and through the **Show in Diagnostics** option. This last option—**Show in Diagnostics**—opens this data in the **Diagnostics** section, with options to further filter the results for analysis.

Since applications are at the heart of ZPA and the end users do not matter if the applications do not exist or if they are down, this dashboard is very important from an administrator's perspective. Let's now explore the second dashboard—namely, the **Users** dashboard.

Users dashboard

On the top left of the **Dashboards** page, clicking the **Users** link (to the right of the **Applications** link) takes an administrator to the **Users** dashboard. This dashboard contains details such as the number of recent users; the number of users currently online and connected to ZPA; top policy blocks for the users; a list of recent users; a list of currently connected users; and a list of users blocked by policies.

The list of users blocked by policies just mentioned can quickly tell us if any users are being blocked by the configured enterprise policy from being able to access certain applications. This could be greatly useful when an end user calls in, reporting that they cannot access a certain application through ZPA.

Positioning the mouse over a user's entry provides a popup with quick details such as the last login time of the user, the application that was accessed by this user, the IP address, and the location of the user. Clicking on the same user entry will open the user in the **Diagnostics** section, with options to further filter the results.

Further down on this dashboard page are widgets that display top users by application, top users by bandwidth, and the top policies by blocked users. This last option—namely, **Top Policies by Blocked Users**—is again helpful for diagnosing whether a certain policy is misconfigured if it is blocking too many end users.

Let's now explore the last dashboard: the **Health** dashboard.

Health dashboard

Clicking on the **Health** link on the top left (to the right of the **Users** link) of the dashboard page takes you to the **Health** dashboard. Monitoring the health of applications from time to time is important for a great user experience and to perform remediation, increase capacity, or simply plan for a downtime for scheduled maintenance.

The **Health** dashboard consists of three sections—namely, **Applications**, **Servers**, and **Connectors**. To reduce the *noise* on this page and keep it less busy and more useful and productive for an administrator, by default only the objects that are either **Down** or **Unhealthy** are visible under these sections.

Application health indicators come from the App Connector. If **Health Reporting** on the App Connector is set to **Continuous**, it reports the health status of all applications within an application segment continuously. If it is set to **On Access**, the App Connector reports the health status only when an end user accesses an application. If it is set to **None**, the App Connector never reports the health status back. The default option is **On Access**.

On the dashboard, different icons in the bottom-right corner of each object indicate their status. Let's look here at what each icon means for an application's health:

- **Up**—A green icon pointing up means that the application is up and functioning.
- **Down**—A red arrow icon pointing down indicates that the application is down and not available for end users. This could be because the underlying server hosting the application is down or unhealthy.

- **Unhealthy**—A yellow icon with a bang (!) means that the application is unhealthy but users can still access it. This could indicate that one or more servers in a multiple server pool could be unhealthy or down. Since there is at least one server available for the application, end users can still access the application.

- **Unknown**—A gray icon with a question mark indicates that the application health is in an unknown state. It could be because health reporting was set to **On Access** and not **Continuous**. When set to **On Access**, ZPA stops reporting on the health of an application if it has been more than 30 minutes since an end user accessed it. In this case, ZPA starts reporting on the application's health again as soon as a user starts accessing the application again.

If an application segment was configured with health reporting set to **On Access** or if an application has not been accessed by a user, then no state is displayed.

There are only two possible health states for App Connectors, as outlined here:

- **Up**—A green icon pointing up means that the App Connector is up and working normally.

- **Down**—A red icon pointing down means that the App Connector is down and not working.

These same states apply to any ZPA **Private Service Edges** (**PSEs**) that an enterprise may have deployed.

Clicking on a server, application, or a connector icon will display a view that can be used to further drill down into the details. Mousing over an icon and clicking on the *map* symbol on the top right of the icon shows a real-time logical connectivity map.

Now that we have looked at the three types of dashboards, let's proceed to the configuration sections of the ZIA Admin Portal. An enterprise administrator will use the **Administration** menu to configure most of the enterprise's ZPA policy.

ZPA administration

The first step for an enterprise ZPA administrator after obtaining the initial credentials from Zscaler is to change the default administrator password and store it in a safe place. The next steps are listed in the following sections.

Company settings

Navigate to **Administration -> Settings -> Company**. Upload an optional company logo and edit the name of the company. This logo and company name will be useful when displaying messages to end users. Clicking the blue **Save** button will save any changes. Note that there is no need to activate any changes made on the ZPA Admin Portal as there is no **Activate** option on the ZPA Admin Portal (unlike the ZIA Admin Portal).

Authentication settings

Next, navigate to **Administration -> Authentication -> Settings**. On this page, click on the **Settings** tab. The first option is to set **Remote Assistance** to **All** or **None**. Setting it to **All** allows Zscaler Support a view-only access to the ZPA Admin Portal. This can be helpful when troubleshooting an issue and Zscaler Support needs to view the settings.

Most enterprise administrators log in to the ZPA Admin Portal using a username and a password. But if an enterprise requires its administrators to log in using a **Security Assertion Markup Language** (**SAML**)-based solution, after adding an **identity provider** (**IdP**) at the **IdP Configuration** page for administrator SSO, the **Enforce SSO Login for Admins** option can be set to **Enabled**.

Type in a value for **Primary Authentication Domain**. This is the principal domain owned by an enterprise. **Additional Authentication Domains** are domains owned by an enterprise that are not configured as primary domains. They can be added only by Zscaler Support upon request. Click on the blue **Save** button to save the changes.

Roles

As with the ZIA administrator roles, the next step is to create the ZPA Admin roles. Click on **Administration -> Settings -> Roles** and go to the **Roles** tab on the page. There are two system-defined in-built roles—namely, the **ZPA Administrator** and **ZPA Read Only Administrator** roles. The former role has full control over all configuration of the Admin Portal, while the latter role just has view-only access to the ZPA Admin Portal and no access to the ZCC Portal. These default roles cannot be changed or deleted.

To add a new role, click on the blue + icon on the top right of the page, and then fill in the following fields:

- **Name**—Enter a role name, such as Helpdesk Associate or Tier 3 Support.
- **Description**—A free-flowing descriptive text field that describes what this role does or who is it for. This shows up on the main **Roles** page, for easy reference when looking at a lot of roles.

- **Access Control**—Under this section, granular control is possible for each of the features such as **Authentication**, **Certificate Management**, and so on. It is recommended that an enterprise put some thought process into the development of these roles, based on the needs and structure of the business.

After making the necessary selections, click on the blue **Save** button to save your changes. To edit any enterprise-created roles, click on the *pencil* icon to the far right of the role under the **Actions** column, and make the necessary changes before saving.

Administrators

Once the various roles have been created as described in the previous section, the next step is to create administrators based on those roles. Click on the **Administrators** tab (to the left of the **Roles** tab) on the top left of the same page.

Unlike the system-defined pre-built roles that cannot be edited or deleted, the default administrator account can be edited or deleted. To create a new administrator, click on the **Add Administrator** link on the top right of the page and fill in the following fields:

- **Admin ID**—This username must be in the form of an email address, and the domain name must match the enterprise domain name explained earlier.

- **Password**—The password must be at least eight characters in length and must include at least one uppercase letter, one number, and one special character.

- **Confirm Password**—Same as for the previous field.

- **Role**—Select the pre-defined role provided by Zscaler, or the custom roles created in the previous section.

- **Status**—If you are creating many accounts and need to activate this account later, set this to **Disabled**. If you want this account to be activated immediately, select **Enabled**.

- **Two Factor Authentication**—Leave this setting to **Off** upon initial creation, and let the administrator first log in to the portal and then go back and enable this feature for their account. An app such as Google Authenticator can be used to enable **two-factor authentication (2FA)**.

- **Force Password Reset**—Enabling this option will force the administrator to change the password upon initial login.

- **Email**—Enter an optional email address for the new administrator.

- **Phone**—Enter a phone number; this is required in case a password recovery is needed in the future.

- **Time Zone**—Select the time zone of where this new administrator is located.

Click on the blue **Save** button to save the changes. Editing an administrator account is as simple as clicking the blue *pencil* icon to the right of the administrator account that is being edited. Make the necessary changes and click on the blue **Save** button.

Certificates

Certificates required for securing ZPA connections can either be created or uploaded to the portal. There are two available options for provisioning certificates to the ZPA infrastructure: the self-contained trust model using self-signed certificates, or the fully trusted model using a subordinate certificate from a trusted external private **certificate authority (CA)**.

To configure these options, navigate to **Administration -> Certificate Management -> Enrollment Certificates**. By default, Zscaler provides a set of CAs on all newly provisioned accounts that can be used right away to enroll the App Connectors and the ZCC app instances on the end-user devices. This set has a **Root** CA and two intermediate CAs generated from it, one for the **connectors** and another for the ZCC app **client** devices.

An enterprise administrator has a choice of whether to generate a completely new set of certificates or to upload certificates signed by their own internal **public key infrastructure (PKI)**. To upload a Root CA certificate or a certificate chain, click on the **Upload Certificate Chain** option, provide a suitable name and description, and click on the **Select File** button. Then, choose the proper certificate file (.pem format) and click on the blue **Upload** button.

An administrator also has an option to generate a certificate by clicking on the **Generate Certificate** option on this page. Provide a suitable name and description, and then select the following options carefully:

- **Type**—This could be a **Root CA** type or an **Intermediate CA** type.
- **Use for ZCC App Enrollment**—The choices are **Yes** or **No**. Selecting **Yes** will use this certificate for the enrollment of the ZCC App for enterprise end users, instead of the Zscaler root certificate.
- **Parent Certificate**—Choose one of the available options; namely, **Root, Client, Connector**, or **CA-Signed-Root**.

After all the options have been selected, click on the blue **Generate** button.

To generate a **certificate signing request (CSR) (CSR)** for an intermediate CA that is to be signed by the enterprise-owned internal PKI, click on the **Create CSR** option. Provide a suitable name and description, and click the **Create** button. This certificate appears in a pending state. Click on the *pencil* icon against it to edit the entry, and then click the **Download .CSR File** option to save the CSR data file and get it signed by the enterprise CA. Once that step is completed, click on the **Select File** option to upload the signed certificate, to activate this intermediate CA.

Let's now look at the steps to configure the ZPA log servers that provide us insights into log activity in real time.

Configuring the ZPA log servers for activity insights

Recall that the log servers for ZPA are very similar to the Nanolog servers we saw for ZIA. However, there is no need for a **Log Streaming Service (LSS) virtual machine (VM)** such as a **Nanolog Streaming Service (NSS)** VM. After logging in to the ZPA Admin Portal, enterprise administrators need to navigate to **Administration -> Log Streaming Service -> Log Receivers**.

Under the **Log Receivers** tab, click on the blue + icon to add a new log receiver, and then fill in the following fields:

- **Name**—Provide a suitable name that can be used to quickly identify the purpose of this log receiver from the main **Log Receivers** page.

- **Description**—A free-flowing text field that can elaborate on what this log receiver does.

- **Domain or IP Address**—Enter a **fully qualified domain name (FQDN)** or an IP address of the receiver.

- **TCP Port**—The port number for the receiver.

- **Connector Groups**—Select one or more of the connector groups already created, and click on the blue **Done** button to save your connector group changes.

Click on the **Next** button to continue with the process. Under the **Log Stream** section, fill in the following fields:

- **Configuration**—Select the log type that is to be streamed. There are five options; namely, **User Activity**, **User Status**, **Connector Status**, **Browser Access**, and **Audit Logs**. Only one option can be selected here.

- **Log Template**—The default setting is **comma-separated values (CSV)** format. Other available options include **tab-separated values (TSV)**, **JavaScript Object Notation (JSON)**, and **Custom**.

- **Log Stream Content**—There are many fields that are included as part of the stream. An administrator can trim some of these as dictated by the business needs so that the log streams do not become too long over time.

- **Policy**—The streamed logs can be filtered right at the source to include just a selected set of session status error codes or a specified segment group. The available options are **SAML Attributes**, **Application Segments**, **Segment Groups**, **Client Types**, and **Session**.

After making the selections, click **Next**. The last section is the **Review** screen. Make sure your choices are what you need before clicking on the **Save** button. If changes are needed, click on the **Previous** button. Clicking the **Cancel** button will discard all changes made so far. Repeat this process for different types of logs, as needed by the enterprise security policy.

Integrating with Azure AD and Okta for SSO

We already discussed the benefits of user authentication in *Chapter 4*, *Understanding Traffic Forwarding and User Authentication Options*. Here, we will look at how to configure ZPA end-user authentication using an IdP such as Azure AD and Okta. The first step in this process is to add an IdP in the ZPA Admin Portal. Enterprise users have an option to authenticate against multiple IdPs.

Adding an IdP

A ZPA administrator needs to first log in to the ZPA Admin Portal and then navigate to **Administration -> Authentication -> Settings**. In the **Primary Authentication Domain** section, the administrator should verify the domains defined for their enterprise. If any domains are incorrect or missing, Zscaler support should be engaged immediately to rectify this situation. Please note that at least one authentication domain per IdP is needed if the administrator plans to add multiple IdPs.

Once this verification is completed, click on the **IdP Configuration** option at the top of the page. Click on the blue + icon in the top right-hand corner of the page to add a new IdP. The resulting popup window presents the following options:

- **Name**—Provide the name of the IdP that you plan you use (such as Okta).

- **Single Sign-On**—Select the **User** option, as we are configuring user authentication. Note, however, that configuring an **Admin** sign-on will require a separate IdP configuration.

- **Domains**—Click on the drop-down list to select the authentication domain you want to associate with this IdP. If multiple domains are selected here, those domains cannot be used in another IdP configuration as a domain can only be associated with a single IdP.

Click **Done** after selecting the domains, and then click **Next**. The next page has the following details:

- **Service Provider SAML Metadata for User SSO**—This section provides the **service provider** (**SP**) metadata that was just generated by the ZPA service. Download the **Service Provider Metadata** (as a file) and, optionally, the **Service Provider Certificate**. Also, this section provides the **Service Provider URL** and **Service Provider Entity ID**.

The preceding information needs to be downloaded by the ZPA administrator and saved in a safe place. If the administrator is not ready for IdP configuration at this time, they should click on the **Pause** button to resume later.

The next step is to configure the IdP. The steps involved in configuring SSO with Okta and Microsoft Azure AD are constantly changing with respect to their **user interface** (**UI**), so please refer to the latest documentation, referenced in the *Further reading* section of this chapter.

Once the IdP configuration steps have been completed, the ZPA administrator needs to return to the **IdP Configuration** page of the ZPA Admin Portal and then click on the IdP that was configured earlier and left in a paused state. On the expanded section of this IdP, copy the **Service Provider URL** and paste it in the appropriate place in the IdP (Okta or Azure AD) configuration page. Perform the same step for the **Service Provider Entity ID**.

Complete the next steps on the IdP portal and then capture the **Identity Provider Metadata** and return to the ZPA Admin Portal. Click on the blue *resume* icon to the right of the IdP (Okta or Azure AD) to resume the process that was paused earlier. On the third page of the **Create IdP** process, upload the IdP Metadata File. Once uploaded, make sure the **IdP Certificate** was also added, and scroll down to set the following options:

- **Status**—If the administrator is satisfied with this IdP configuration, set this option to **Enabled**.

- **ZPA (SP) SAML Request**—Set this option to **Signed** if the administrator wants outgoing requests to the IdP to be signed.

- **HTTP-Redirect**—Set this option to **Enabled** if the IdP needs a **HyperText Transfer Protocol (HTTP)** redirect instead of a POST method.

Clicking the **Save** button after verifying the **Authentication Domains** will finalize this IdP. To quickly view the details for a configured IdP, click on the *chevron* icon to the left of the IdP name; to edit the configuration for an IdP, click on the *pencil* icon to the right of the IdP name; and to delete the IdP configuration altogether, click on the red *x* icon to the right of the IdP name.

Finally, verify the user authentication using this IdP is working as intended, and import the SAML attributes as per the needs of the enterprise. These SAML attributes can be viewed by clicking the **SAML Attributes** tab at the top of the page, right next to the **IdP Configuration** tab. These attributes can then be used to configure access for end users, using policy configurations.

Configuring the ZCC app for ZPA

Let's now learn how to prepare the ZCC app for a ZPA deployment within an enterprise. We already discussed the options supported by the ZCC app in detail in the context of ZIA, in *Chapter 4, Understanding Traffic Forwarding and User Authentication Options*, so we will not duplicate them here and only mention ZPA-specific options.

ZCC app installation

The first step is the installation of the app by the appropriate IT department within an enterprise. A few settings can be set beforehand for groups of enterprise users that will be using the app. Those settings include **App Profile**, **Forwarding Profile**, **Notification**, **Support**, **Trusted Network**, **Zscaler Service Entitlement**, **User Agent**, and **Device Posture** configurations.

On mobile devices, the ZCC app can be installed using the applicable app stores by the end users themselves. In a large enterprise, the ZCC app can be deployed using a **mobile device management (MDM)** platform. A MDM platform allows an administrator more control over the installation and configuration settings, saving additional effort for end users.

ZCC app enrollment and authentication

Once the app is installed on the end users' computers and mobile devices, the end users then enroll into the app. The enrollment also authenticates the end users to the ZPA service and may be prompted to authenticate at periodic intervals controlled by the administrator, using the **Timeout Policy** settings on the ZCC Admin Portal. If the app is also used for ZIA authentication, it is recommended to use the same SAML IdP to avoid confusing an end user into authenticating twice, once for ZIA and once again for ZPA, using different IdPs.

ZPA application access

Once an end user is enrolled successfully for ZPA and requests an application supported by ZPA (for which the end user is allowed access per the configured policy), the ZCC app will establish an encrypted **ZPA tunnel (Z tunnel)** to the closest healthy ZPA PSE. ZPA then brokers an end-to-end microtunnel to the appropriate App Connector, thus establishing a data path between the originating end user and the destination application.

There are three important traffic-forwarding scenarios for the ZCC App—Off Trusted Network, On Trusted Network, and VPN Trusted Network.

Off Trusted Network

In this scenario, an end user is not connecting from an enterprise location. Examples include the user's home, an airport, or even a location with a public Wi-Fi network. User traffic is tunneled through the ZCC app, and this option can be potentially used as a replacement for traditional **virtual private networks (VPNs)**.

On Trusted Network

When enterprise users connect from the head office or a regional office, they are said to be coming from a trusted location. In this scenario, as the trusted location already has secure connectivity to the enterprise private applications, there is no need for the ZCC app to tunnel the traffic. However, enabling tunneling in this case allows ZPA to offload access to cloud-based applications and helps reduce bandwidth consumption between the data center and the cloud applications.

VPN Trusted Network

In this scenario, an end user establishes a VPN connection to the trusted network from a non-enterprise location. A full-tunnel VPN carries all traffic over the VPN. A split-tunnel VPN carries only specific traffic over the VPN and offloads the remaining traffic over the end user's local internet breakout.

Since ZPA is positioned to be a replacement for VPN, when running the ZCC app alongside a VPN careful configuration is necessary to avoid any conflicts. Zscaler recommends a best practice of disconnecting any VPN tunnels when ZPA is active.

Device posture control

Each end-user device can be configured with a Posture Profile. This is a set of criteria that the device must meet before it can be allowed to access ZPA applications. These posture profiles can be added per platform, such as Windows, macOS, iOS, and Android.

ZPA process flow

Let's review the high-level steps involved in the end-to-end ZPA process flow, as follows:

1. An enterprise administrator configures the ZCC app settings for end users or groups in App Profiles, and—optionally—one or more Forwarding Profiles.
2. The IT teams in the enterprise distribute and install the app to end users.
3. The end users then enroll and authenticate into the app.
4. Upon successful enrollment, the app gets its matching profiles, including the identity certificate needed for authenticating the Z tunnels. The app is also notified of the available private applications by the ZPA PSE.
5. The app sends the host device information to Zscaler to allow for device fingerprinting, to prevent possible cloning of the machine for unauthorized access, thereby guaranteeing security.
6. When the end user requests access to an enterprise private application, the app sends that request to the ZPA cloud via the ZPA PSE.
7. The ZPA PSE then replies to the app, and the app allocates a synthetic IP address from the `100.64.0.0/16` range for that application.
8. The ZCC app and the App Connector adjacent to the application will establish outbound **Transport Layer Security** (**TLS**) 1.2-encrypted tunnels to the ZPA PSE.

9. The app then sends the data from the source to the microtunnel established for this connection. This same microtunnel connects at the ZPA PSE to the microtunnel established to the connector, providing an end-to-end connection from ZCC app to the App Connector.

In summary, the browser or the software agent on the host device believes that it is talking directly to the ZCC app, and the application believes that it is talking directly to the connector. The data flow between the endpoints is intelligently managed by the ZCC app and the App Connectors.

Summary

In this chapter, we started our journey toward configuring the various components of a ZPA solution for an enterprise. We started with an overview of the ZPA Admin Portal, configured ZPA log servers and user authentication, and ended with in-depth configuration settings for the ZCC app.

In the next chapter, we will continue our journey and complete the remaining configuration of the ZPA admin portal. We will also learn how to integrate ZPA with the IdP for SSO, look at best practices for enterprise deployments, and deploy connectors in on-premises data centers and the public cloud.

Questions

As we conclude, here is a list of questions for you to test your knowledge regarding this chapter's material. You will find the answers in the *Assessments* section of the Appendix:

1. ZPA applications should be ideally configured with the **Dynamic Server Discovery** option.

    a. True

    b. False

2. Every new ZPA account comes with the following default certificates:

    a. Root

    b. Connector

    c. Client

    d. All the above

3. **Secure Sockets Layer (SSL)** interception can be performed on the path between:

   a. The ZCC app and the ZPA cloud

   b. App Connectors and the ZPA cloud

   c. Neither of the above

   d. Both of the above

4. Applications such as an active **File Transfer Protocol (FTP)** and **Voice over IP (VoIP)** work great on ZPA.

   a. True

   b. False

Further reading

- Configuring an IdP for SSO: `https://help.zscaler.com/zpa/ configuring-idp-single-sign`
- Configuration guide for Okta: `https://help.zscaler.com/zpa/ configuration-guide-okta`
- Configuration guide for Microsoft Azure AD: `https://help.zscaler.com/ zpa/configuration-guide-microsoft-azure-ad`

9
Using ZPA to Provide Secure Application Access

In this chapter, we will continue our journey into the ZPA Admin Portal configuration. Specifically, we will learn about the best practices when it comes to deployments within an enterprise, and the steps involved in deploying the App Connectors in both on-premises environments and cloud accounts.

These topics will further solidify the understanding you'll need to deploy a ZPA solution within an enterprise as an administrator. In this chapter, we are going to cover the following topics:

- Deploying App Connectors
- Configuring ZPA applications
- Exploring the best practices for enterprise deployments

Deploying App Connectors

Now that we've reviewed how to install and configure the ZCC app for the originating enterprise end users, let's learn how to deploy the App Connectors for the enterprise's private applications.

Connector requirements

App Connectors are the only elements of ZPA that connect to the enterprise's internal network, near the private applications that need sharing. A **connector** is a lightweight Linux-based implementation that boots up extremely quickly and provides access to applications. Connectors only establish outbound connections to the ZPA infrastructure using a provisioning key.

Upon bootup, the connector is configured to contact the nearest healthy ZPA PSE. A control channel is then established that enables us to register and configure the connector. This also allows the connector to inform the ZPA **Central Authority (CA)** of the applications that have been discovered. Although not a requirement, it is recommended to deploy connectors in pairs for high availability in case one connector is down or undergoing an upgrade.

A **Virtual Machine (VM)** infrastructure is necessary for deploying a connector on-premises. Some of the supported on-premises environments are CentOS, Oracle Linux, Red Hat Enterprise Linux, and VMware variants such as VMware Center and vSphere Hypervisor (ESXi). The supported cloud platforms include Amazon Web Services and Microsoft Azure.

The next requirement is a valid connector provisioning key that is available from the ZPA Admin Portal. To connect to the outside network, connectors also need IP addressing (dynamic using DHCP or a static IP) and DNS resolution, capable of resolving both internal and external hosts. Connectors must use static **Media Access Control** (MAC) addresses and have internal network connectivity to the application servers.

If the enterprise deploys a firewall or restricts outbound internet traffic from the data center, the firewall must be configured to allow outbound connectivity on port 443 to perform **Network Address Translation** (NAT) for the connector source IP addresses.

To avoid interoperability problems with other security products and services, including Zscaler Internet Access, it is recommended that you don't send the connector traffic through an existing Zscaler tunnel (GRE or IPsec) or through any encapsulation that may interfere with the use of a standard 1,500 byte **maximum transmission unit** (MTU).

Because ZPA enforces TLS certificate pinning for both the client and server certificates, all forms of inline or man-in-the-middle TLS interception or inspection must be disabled. If the TLS certificates presented by ZPA PSEs cannot be cryptographically verified against Zscaler-trusted public keys, the App Connectors will not function. This certificate verification process cannot be reconfigured by design to maintain the integrity of the entire system.

Each connector can support a throughput of up to 500 Mbps and although they can scale vertically, it is recommended to implement horizontal scaling. If all your applications plan on using the double encryption feature, this throughput is reduced to 250 Mbps. The minimum hardware requirements include 4 GB RAM, two Xeon E5 class CPU cores (four cores with hyperthreading for VM connectors), 8 GB disk space (thin provisioned), and at least one **Network Interface Card** (**NIC**). If ZPA LSS is being used, it's recommended that you deploy additional connectors adjacent to the enterprise SIEM, just for this streaming option. This avoids any possible contention between end user traffic and log streaming traffic.

Installing the connector

The first step is to provision the VM as per the requirements mentioned previously. Upon booting up for the first time, the connector does not have its own key pairs, so it proceeds to generate its own public/private key pairs. The private keys are encrypted using a hardware fingerprint and stored locally and never shared with external entities.

The connector then proceeds to generate a **certificate signing request** (**CSR**) using its private key, which is stored locally. The connector authenticates this CSR to the ZPA cloud using the provisioning key that was previously obtained from the ZPA Admin Portal.

The ZPA cloud eventually issues a signed TLS client identity certificate and a signed server certificate to the connector. These certificates are pinned to the hardware fingerprint of the connector and are attached to a single enterprise account. This prevents connector duplication in VM environments as the keys will not match the virtual hardware fingerprints.

It is not recommended to register a connector through a proxy, although this can be done as needed, such as in a *"no default route"* environment. If a proxy is indeed in place, then it should not perform request authentication from the connectors and should not do any form of inline inspection. This will prevent the connectors from registering and establishing the Z tunnels to the ZPA PSE.

Connector updates

The connector groups are usually configured with a 4-hour update window, which will be used to automatically perform software updates on the connectors that are part of that group. Connectors within the group are selected at random and upgraded one at a time. Before the upgrade begins, the connectors drain existing connections and do not accept new connections for 5 minutes.

During the upgrade, the connectors do not accept any new connections and those new connection requests are redirected to other available connectors within the group. Once the upgrade is successful, the connector resumes accepting application requests. Another randomly selected connector within the group is then chosen and upgraded in a similar fashion. This process continues until all the connectors within that group are upgraded.

A manual upgrade option for the connectors is also available via the ZPA Admin Portal. It is recommended that this is done immediately after deploying a new connector. The requirement is that the connector indicates a status of **Scheduled** on the portal. The administrator can click on the **Update Now** option to initiate this manual upgrade. Another way to initiate the manual upgrade is to type in the `set sudo yum update -y` and `sudo reboot` commands from the connector CLI.

Connector provisioning

Now, let's learn how an administrator can provision a new connector on the ZPA Admin Portal. After logging into the ZPA Admin portal, the administrator needs to navigate to **Administration -> Connector Management -> Connectors**.

On the top right-hand side of the page, click on the blue + icon to start adding a new connector. From the pop - up window, select the radio button that says **Create a new provisioning key** and click **Next**.

On the second leg of the process, click on the drop - down list to **Choose a certificate**. Select the radio button that says **Connector** as this certificate is for a connector. Click **Next** to continue with the process.

On the third leg of this process, click on **Add Connector Group**, which shows the following options:

- **Name**: Specify a name for this connector group.

- **Description**: A description for this connector group.

- **Status**: If you are creating a bunch of connector groups for future use, you can leave this option **Disabled**. If you plan to use this connector group right away, select **Enabled**.

- **Connector Software Update Schedule**: Select the upgrade window according to your enterprise policy regarding software updates. Choose a day of the week and the start time using a 24-hour clock.

- **Connector Location**: Type in your location using the city, state, and country notation. This makes sure the upgrade window specified previously matches the proper time zone.

Click **Next** to continue. On the **Create Provisioning Key** tab, set the following options:

- **Name**: The name of this provisioning key.

- **Maximum Reuse of Provisioning Key**: This number specifies the number of times this provisioning key can be used to register an app connector.

- **Instances of Provisioning Key Reuse**: This field cannot be modified by the administrator. ZPA automatically keeps a count of the App Connectors that have been registered in this app connector group and shows that number in this field.

Click **Next** to continue with this process. Verify all the information you've entered so far on the **Review** screen and click the **Save** button to add this connector configuration. The next screen shows the actual provisioning key. Click on the copy icon beside the **Copy Provisioning Key** option to copy the key to clipboard, save it to a Notepad file, and keep it in a secure location. Choose your platform under the **Review Documentation** section to get the app connector deployment instructions for that platform. Finally, click the **Done** button to complete the **Add Connector** process.

Please note that on the **Connectors** main page, this connector will not show up yet because it has not been installed and activated. On the same page, click on the **Connector Groups** tab to view the list of available connector groups. Clicking on the **Connector Provisioning Keys** tab displays a list of the provisioning keys for the connectors.

The deployment guides for VMware and cloud platforms are referenced in the *Further reading* section of this chapter. Now that we understand how to deploy the App Connectors, let's focus on how to configure private applications.

Configuring ZPA applications

An application is defined as a combination of a port, along with a **Fully Qualified Domain Name** (FQDN), IP address, or a local domain hostname.

The next step after deploying the App Connectors is to configure the private applications themselves. After authenticating into the ZPA Admin Portal, the enterprise administrator needs to navigate to **Administration -> Application Management -> Application Segments**. This page displays a list of the existing application segments.

DNS search domains

If the enterprise wants end users to be able to access the private applications using a short name rather than the FQDN, the administrator needs to add the relevant domains that can be used to form an FQDN. For this, the administrator needs to select the **DNS Search Domains** icon from the top right-hand corner of the **Application Segments** page.

In the pop - up window, enter the domain name and then click on the **Add More** link to add more domains on a new line. This process can be repeated for as many domains as needed. There is a checkbox to the right of each domain name that says **Domain Validation in Zscaler App**. Enabling this option allows the ZCC app to try and resolve these domains first, ensuring proper resolution and reachability of the ZPA applications. This also helps the ZCC app by preventing DNS responses being hijacked by some service providers that use DNS optimization techniques, which could cause the ZCC app to incorrectly resolve the synthetic IP address for a ZPA application, making them appear as unreachable. Click **Save** to confirm your chosen options.

Adding an application segment

A group of defined applications (based on user privileges or your type of end user access) is called an **application segment**. ZPA features such as health reporting and double encryption are defined per application segment.

On the **Application Segments** page, click on the blue + icon in the top right-hand corner of the page to add a new application segment. On the first **Define Applications** tab, enter the following information under the **General Information** section:

- **Name**: Enter the name of the application segment.
- **Description**: A short description of this application segment.
- **Status**: If this application segment will be used right away, choose **Enabled**; otherwise, set this to **Disabled**.

Under the **Applications** section, enter one or more application identifiers, such as a single or a wildcard FQDN, a hostname, an IP address, or a subnet. If browser access is being used, then select the checkbox that says **Browser Access** to the right of the application identifier. Enabling the **Browser Access** option requires further configuration, which is outside the scope of this book. You also have the option to add more application identifiers by clicking on the blue + icon that says **Add More**.

Under the **Zscaler Client Connector Access** section, enter the **TCP Port Ranges** and the **UDP Port Ranges** details for your enterprise application. The **Add More** option can be utilized to add more than one range for each option.

Under the **Additional Configuration** section, set the **Double Encryption** option to either **Enabled** or **Disabled** based on your enterprise needs. Set the **Bypass** option to **Use Client Forwarding Policy** (the default option, which equals **Never**, **Always**, or **On Corporate Network**). Enabling this **Bypass** option allows the end users to bypass ZPA to access an application under those selected conditions.

Under the **Common Configuration** section, **Health Reporting** can be set to either **Continuous** or **On Access**, depending on your enterprise policy. The options for **Health Check** are **Default** or **None**. The **Zscaler Client Connector can receive CNAME** option can be **Enabled** or **Disabled**. When enabled, this allows the App Connectors to resolve CNAME records.

Once you've finished adding the necessary application identifiers, click **Next** to continue. Under the second **Segment Group** tab, click on **Add Segment Group** and provide a **Name** and **Description** for this segment group. To start using this segment group immediately, set **Status** to **Enabled**; otherwise, set it to **Disabled**. Click **Next** to continue.

On the third **Server Groups** tab, click on **Add Server Group** and fill in the following options:

- **Name**: Enter the name for this new server group.

- **Description**: Type in the description for this new server group (for example, web servers).

- **Status**: This can be set to **Enabled** if you wish to start using it immediately; otherwise, choose **Disabled**.

- **Dynamic Server Discovery**: To allow ZPA to discover servers automatically when end users request application access, set this option to **On** (the default and recommended option). If this setting is turned **Off**, then additional manual configuration will need to be added for the servers hosting the applications.

- **Connector Groups**: Select one or more connector groups that were configured previously.

Click **Done** and then **Next** to continue with this configuration process. On the **Review** tab, verify all your settings for accuracy and then click **Save** to finish adding a new application segment. On the last **Policies** tab, the administrator has the option to add access control options by clicking on the **Edit Policy** tab. To do this task later, click **Cancel**.

Now that we have finished adding applications, the next step is to configure access control policies for the end users.

Configure SAML attributes

SAML attributes are simply properties for an end user such as their first name, last name, email address, and department name. Before we can assign application access based on these attributes, we first need to obtain the necessary attributes. We learned how to set up the SAML authentication with popular IdPs at the beginning of this chapter. Now, let's look at how to configure the SAML attributes for end users.

While logged into the ZPA Admin Portal, navigate to **Administration -> Authentication -> SAML Attributes**. Click on the **SAML Attributes** tab at the top of the page to view the SAML attributes for all the IdPs that have been configured already. Click on the **All** option under **IdP Configuration** at the top of the page and select the appropriate IdP (for example, ADFS).

Click on the blue + icon in the top right-hand corner of the page to add a new SAML attribute. Fill in the following options on the resulting pop - up window:

- **Name**: Enter an appropriate attribute name.
- **SAML Attribute**: Select the matching SAML attribute.
- **IdP Configuration**: Click on the drop - down list under this option and select the radio button for the IdP.

Click **Save** to add this SAML attribute for the selected IdP. Now that we have the attributes for end users, the next step is to create access policies for them based on these attributes.

Configuring end user access policies

To start configuring the end user access policies, the administrator needs to navigate to **Administration -> Policy Management -> Access Policy**. Because ZPA follows the **zero-trust network access (ZTNA)** principles, by default, there are no access policies configured under the **Access Policy** tab. This means that by default, all users are denied access to all applications. To allow an end user access to an application, a new access policy rule must be created to explicitly allow that end user to access that application segment or a segment group. A segment group is a collection of similar application segments. One application segment can only be part of one segment group.

Click on the blue + icon in the top-right corner to start adding a new access policy rule:

- **Name:** Enter the name for this access policy; for example, `Allow Payroll Access`.

- **Description**: Provide a suitable description, for example, `This access policy rule allows HR personnel to access the Payroll system`.

Under the **Action** section, provide the following information:

- **Rule Action:** If this is an access policy that's explicitly allowing access, then select the **Allow Access** option. If this access policy explicitly denies access (for example, non-HR personnel are explicitly denied access to the payroll system), then select the **Block Access** option.

- **App Connector Selection Method**: There are two options available, namely, **All App Connector Groups for this application** and **Specific App Connector or Server Groups for the application**. Your selection depends on the preference of your enterprise.

- **Message to User**: An optional free-flow text message can be entered here that will be displayed to the user that matches this access policy rule. If this is a block access policy, a message may be displayed to the user that they are being denied access to this application, so the user is more informed and not wondering what happened. For example, the message could say **Access to company payroll systems is only allowed for HR department personnel**.

The next **Criteria** section can get a little complicated. The options here are as follows:

- **Application Segments**: Choose one or more application segments that were already defined and click **Done**.

- **Segment Groups**: Pick one or more segment groups defined previously and click **Done**.

Note that the application segments and the segment groups that have been selected are combined using logical OR logic.

- **SAML and SCIM Attributes**: Click on **Select IdP** to show a list of the configured IdPs. Select one of them; the attributes for that IdP will be loaded underneath. Click on the **Any SAML Attribute** drop - down list and choose one of the attributes for the user, such as first name, last name, or department name. Click on the **Add More** option to select more SAML attributes.

- **Client Types**: Click into the **Any client type** drop - down list to select either **Client Connector**, **Machine Tunnel**, **Web Browser**, or **ZIA Service Edge**. A machine tunnel allows an end user Windows device to establish connectivity to ZPA even before the end user is logged into the ZCC app.

- **Machine Group**: Machine groups are local, internal enterprise machines that need ZPA connectivity. You can select a configured machine group here if necessary.

- **Client Connector Posture Profiles**: The options here are **Verified** and **Verification Failed**. The first option means that the ZCC app has been verified using the posture settings that have been configured in the ZCC portal. The second option indicates that the configured posture settings have failed.

- **Client Connector Trusted Networks**: Choose one or more trusted networks that have already been configured. These trusted networks are joined using a logical OR combination.

Click **Save** to create this access policy. Please note that the logical evaluation of all these settings is as follows:

```
(Application Segments) OR (Segment Groups)
AND
SAML and SCIM Attributes
AND
Client Types
AND
Machine Group
AND
Client Connector Posture Profiles
AND
Client Connector Trusted Networks
```

Since the access policy evaluation contains complex evaluation logic, please exercise care in creating the proper policies on paper before you start to create these policies in the ZPA Admin Portal.

Once these access policy rules have been created, they will appear on the **Access Policy** tab. It is a best practice to place the most specific access policy rules at the top and the least specific access policy rules toward the bottom. The access policy rules can easily by moved around by an administrator by clicking on the access policy's **Rule Order** number. An option will appear, allowing them to edit the rule order number. For example, you can click on rule number 2 and then simply change it to rule number 3 and the rule will automatically move down to the third place.

ZPA timeout policy

A ZPA timeout policy allows you to configure the timeout values for authentication and idle connections. Click on the **Timeout Policy** tab at the top of the page.

ZPA has a single, default timeout policy. Click on the *pencil* icon to the right of the default rule to edit it. The name and description of the default rule cannot be edited. Under the **Timeouts** section, you have the following options:

- **Authentication Timeout**: The default option is set to **Specific Interval** and that interval is 7 days. This number can be lowered, and the **Days** dropdown field can be changed to **Hours** or even **Minutes**. It can also be set to **Never** if the enterprise does not want the end users to authenticate periodically.

- **Message to User**: If the authentication timeout is set to a certain value in days, hours, or minutes, a message can be displayed to the end user. For example, it may say **Your session has expired, please authenticate again.**

- **Idle Connection Timeout** – The default is set to **Default**. This means that the end user sessions never end, even after periods of inactivity.

Under the **Criteria** section, you can make changes as needed. Click **Save** to apply the changes to the default timeout policy or click **Cancel** to discard any changes. Back on the main page, you can click on the blue + icon in the top right-hand corner of the page to create your own timeout policy for the entire organization, or even just for a handful of privileged users that will override the default timeout policy. For example, you may want to expire administrator sessions after 1 hour to prevent any potential security risks.

Now that we've looked at how to grant end users access to applications, let's look at available options to check on the health of these applications.

Application health monitoring

By default, ZPA only checks on the health of its applications when the end users are accessing them. If an administrator manually defines the enterprise applications instead of using automatic discovery, the available health check option is called active or continuous.

The **Active health monitoring** option causes the App Connectors to poll each application instance on one port every second, and then continue to check the remaining defined applications and ports in a round-robin fashion. This means the larger the application list and the number of ports, the longer it takes for one complete health check cycle to be done through all of them.

If the administrator enables application discovery, then ZPA defaults to the default behavior explained previously, where it only checks the health of the application when end users are accessing it, and up to 30 minutes after that initial access time.

Active health monitoring is only available for applications that have been defined with specific hostnames and not for applications with a wildcard domain or an IP address. There is also a maximum port count of 10 per application. We looked at these options earlier in the *Adding an application segment* section.

Now, let's explore the best practices for deploying ZPA in enterprises.

Exploring the best practices for enterprise deployments

The best practices for an enterprise ZPA deployment can be divided per component. Let's start with the connectors.

App Connectors

When configuring your App Connectors as part of an enterprise deployment, you should install the App Connectors on an internal network segment, adjacent to the private applications. It is recommended that the network segment be configured with a default route to the internet.

The connector should also have access to an internal DNS server that can resolve both the application hosts and the hosts on the internet. Full application port and protocol access, including **internet control message protocol (ICMP)** access, should be granted to the connectors. If possible, avoid an explicit proxy on the path between the connector and the ZPA cloud. ICMP access is required for **User Data Protocol (UDP)** applications, which allows the **round-trip time (RTT)** calculations to the application host. This is needed if you wish to make load balancing decisions for the end user traffic.

If possible, avoid an explicit proxy on the path between the connector and the ZPA cloud. You should also disable any SSL interception on this path as well to properly establish the connection between the connector and the ZPA cloud.

Each connector group is recommended to have at least two connectors for high availability, in case one connector goes down due to a failure or due to a software upgrade. Try to meet the minimum resource specifications for the virtual machine to be able to reach the specified throughput and capacity.

It is also recommended to deploy connectors with an N+1 redundancy; for example, if an application needs 2 Gbps throughput. In such a case, four connectors (500 Mbps x 4 = 2 Gbps) should be sufficient during normal operation. However, if one connector were to go down due to a failure or a software upgrade, then there will only be three connectors with a 500 Mbps x 3 = 1.5 Gbps throughput. So, in summary, N = 4 here, so it is recommended to deploy 4+1 = 5 connectors instead.

Certificates

By default, when a ZPA account is provisioned for an enterprise, it comes with a complete set of self-signed root, client, and connector certificates on the Admin Portal under **Enrolment Certificates**. These certificates can be used immediately by the enterprise.

However, if the enterprise chooses to use its own certificates or plans to use the double encryption feature of ZPA, then they will need to get those certificates signed by their own internal private root **CA** and upload them to the ZPA Admin Portal. Please note that in this case, the enterprise needs to manage the certificate life cycle, such as creation and expiry.

Authentication

For end user authentication, it is best to use the same IdP with SAML for both ZIA and ZPA authentication to avoid users logging into two separate IdPs and causing confusion. For administrator authentication, it is recommended to use **single sign-on** (**SSO**) with a supported IdP, preferably with **multi-factor authentication** (**MFA**).

ZCC app

Like the App Connectors, it is recommended to not have a proxy between the end user device running the ZCC app to the ZPA infrastructure. Also, attempting to perform SSL inspection on the same path will fail to establish ZPA connectivity.

To reduce the burden on the end users, the enterprise IT department should silently deploy the app and provide the domain and the app profile during that installation. This allows the end users to log in just with their username instead of their email address. The ZPA administrator should also work with the IT department to make sure the correct app profile and related settings are automatically attached to the app.

If the ZCC app is used for both ZIA and ZPA, it is better to use SAML with the same IdP to avoid users needing to authenticate twice. Optionally, MFA can be added for ZPA authentication, which adds an additional layer of security to private application access.

Application

When creating ZPA server groups, it is recommended to use the **Dynamic Discovery** option for the servers. This allows ZPA to automatically select the best server instance for each end user request. Use a static option to define servers to limit the number of IP addresses that can serve a hostname request, or even to override the end user request by routing to an IP address of your choice.

Configure your application segments properly to avoid a poor end user experience. If an application segment is associated with multiple valid connector groups, ZPA will select the nearest connector group to the end user based on the geographical location of the client and the geographic location of the connector group.

For example, if an enterprise has offices on both the US east coast and the US west coast, make sure the connector groups are properly configured based on their geographic location. A misconfiguration could send end users on the east coast to the connector groups on the west coast!

Applications such as internal enterprise DNS are not supported by ZPA. If an end user requests an application using an FQDN, the internal DNS resolves this to an IP address. The end user will then try to access the application using this IP address instead of going to ZPA. Similarly, the suite of unified communications such as **Voice over Internet Protocol (VoIP)** and video are unsuitable because they use peer-to-peer mechanisms.

In active mode **File Transfer Protocols (FTPs)**, the FTP server calls the originating end user client on a random port to establish a data connection. Since the server is not aware of this Mtunnel architecture, it will not be able to route back to the client.

Monitoring

It is recommended to leave the **Health Check** option for an application segment set to **Default** and the **Health Reporting** option set to **On Access**. This setting ensures that the health of the application is monitored for 30 minutes after the initial user access request. If the applications do not take kindly to unsolicited health check requests where those health check requests consume the listening socket on the server, set the **Health Check** option to **None**.

Log streaming service

In an enterprise production setting, it is recommended to install the LSS connector infrastructure adjacent to the SIEM and add them to a dedicated connector group receiving the logs. If the connector group is shared with user traffic, a rush of user traffic could impact the delivery of the logs, causing the potential loss of those logs.

Keeping redundancy in mind, it is also recommended to stream LSS logs to at least two log receivers at geographically separate locations and use separate connector groups for each log receiver destination. This is very much like the NSS recommendation we saw for ZIA.

Summary

In this chapter, we continued our journey toward configuring the various components of a ZPA solution for an enterprise. We learned the steps involved in provisioning App Connectors and the applications themselves. We also explored the best practices that an administrator should know about before attempting to deploy a ZPA solution within their enterprise.

In the next chapter, we will explore the migration process to ZPA and how to troubleshoot the most common ZPA end user issues.

Questions

As we conclude, here is a list of questions for you to test your knowledge regarding this chapter's material. You will find the answers in the *Assessments* section of the *Appendix*:

1. Every new ZPA account comes with which of the following default certificates?

    a. Root

    b. Connector

c. Client

d. All the above

2. SSL interception can be performed on the path between which of the following?

a. The ZCC app and the ZPA cloud

b. App Connectors and the ZPA cloud

c. None of the above

d. All the above

3. Applications such as active FTP and VoIP work great on ZPA.

a. True

b. False

Further reading

- App Connector Deployment Guide for VMware Platforms: `https://help.zscaler.com/zpa/connector-deployment-guide-VMware-platforms`

- App Connector Deployment Guide for Amazon Web Services: `https://help.zscaler.com/zpa/connector-deployment-guide-amazon-web-services`

- App Connector Deployment Guide for CentOS, Oracle, and Red Hat: `https://help.zscaler.com/zpa/connector-deployment-guide-centos-oracle-and-redhat`

- App Connector Deployment Guide for Microsoft Azure: `https://help.zscaler.com/zpa/connector-deployment-guide-microsoft-azure`

- App Connector Deployment Guide for Microsoft Hyper-V: `https://help.zscaler.com/zpa/connector-deployment-guide-microsoft-hyper-v`

- Machine Tunnels: `https://help.zscaler.com/zscaler-client-connector/about-machine-tunnels`

10
Architecting and Troubleshooting Your ZPA Solution

So far, we have seen the various steps involved in configuring a ZPA solution. In this chapter, we will start from the existing application access landscape within the enterprise and plan our migration to a ZPA solution.

We will step through the Zscaler Question Set for the ZPA solution. Through this question set, we will gather all the requirements for the ZPA solution and then start building it and migrate the enterprise users to that solution.

We will then look at the various scenarios that can be troubleshooted effectively and efficiently using a guide that can be used by the ZPA operations team of the enterprise.

In this chapter, we are going to cover the following topics:

- Architecting your ZPA solution
- Troubleshooting your ZPA solution

Architecting your ZPA solution

Before the existing enterprise applications can be migrated to ZPA, a methodical approach is needed to classify those applications and map the end user's access to those applications. Each enterprise needs to perform this task in their own way. The ZPA administrator needs to coordinate the discovery of all the applications that need to be migrated to ZPA.

First, obtain the list of applications that need to be migrated to ZPA. A standard questionnaire can be developed per the enterprise guidelines, and all the collected data needs to be classified accordingly.

The next step is to create application segments and then add the necessary segment groups and server groups. The application connectors could reside in on-premises data centers or in public cloud environments such as AWS and Azure.

The final step is to map the end user access to these applications. Remember that users are denied ZPA access to every private application by default and that access must be explicitly configured. This access could be based on the user's department, job title, location, and so on. This needs to be modeled after the enterprise access policy.

These steps are just at a high level. The ZPA Question Set helps a ZPA administrator bring all the deployment details under one roof. Let's explore it in more detail.

Stepping through the ZPA Question Set

Let's review the various steps involved in the ZPA Question Set. Navigate to the main page of the Zscaler Question Set and click **Yes** to start the wizard. We will skip the fields on the first page because this was already covered in *Chapter 5, Architecting and Implementing Your ZIA Solution*.

Select the **Zscaler Private Access (ZPA)** checkbox and click **Next** to start the process for a ZPA solution.

Approximate number of users planned to be on Zscaler

There are two main divisions under this section. The first calls for the total number of users, while the second calls for the number of remote users:

- **Approx Number of Total Production Users (Including Guests & Contractors)?:** The total number of enterprise users that will be using the ZPA solution. This number should include any contractors, guests, vendors, and so on, in addition to the employees of the enterprise.

- **Approx Number of Remote Users (Road Warriors)?**: Zscaler calls remote and traveling users as road warriors. These users do not typically work out of a fixed office location.

Click **Next** to continue the process.

Package management

The **Zscaler Client Connector (ZCC)** app is a central component for the end user to access private applications in ZPA. Hence, the next question is how the enterprise plans to deploy the application to the end user devices.

For the main question, **How are Application Packages Deployed for the Following Managed Devices?**, the platforms listed are **Windows, MAC OSx, GNU/Linux, iOS Devices**, **Android Devices**, and **Chrome OS**. There is a free-flowing text field at the end that captures any other platforms not already listed. Click **Next** to continue.

Antivirus/security solutions

The ZCC app will be deployed to the various platforms mentioned in the previous section. To prevent any antivirus, firewall, or endpoint security solutions from interfering with the functionality of the ZCC app, it is important to document which solutions are in use for each platform. The same platforms listed in the previous section are listed here.

Click **Next** to move on with the process.

Remote access VPN deployment

Recall that ZPA is positioned to replace the existing VPN solution for an enterprise. Hence, it is important to find out if there is an existing VPN solution and how it works:

- **Which VPN Clients are Supported in your Environment?**: Choose the VPN client that is in use within the enterprise. Some of the popular options are **Cisco AnyConnect** and **Juniper Pulse Secure**. There are other options available in the dropdown list.

- **How is Remote Access VPN Connection Established?**: End users can establish the VPN connection in two ways. The options are **On Demand** and **Always On**.

- **How is Remote Access VPN Deployed?**: The first option is **Split-tunnel (Best Practice)**. This means that only the enterprise-chosen traffic is routed over the VPN client; the rest of the end user internet traffic goes over their local internet breakout connection. The second option is **Split Tunnel and Split DNS**. In this case, traffic over the VPN client uses the DNS server specified by the VPN client; the rest of the traffic uses local DNS servers. The last option is **Full Tunnel**, and is where all the traffic goes over the VPN client.

- **Can the Internal DNS Servers Resolve External Public Domains?**: There are only two options: **Yes** or **No**. Recall our ZCC app interaction with DNS servers and its implications in *Chapter 8, Exploring the ZPA Admin Portal and Basic Configuration*.

Click **Next** to carry on with the process.

User management

This section evaluates the various available options for end user authentication:

- **Please Select Current SAML 2.0 Federation Vendor**: If the enterprise already uses a SAML 2.0 compliant vendor, then it is possible to use the same vendor for user authentication. Various popular vendors are available in the drop - down list to choose from.

- **Can SAML 2.0 Federation Authenticate All Required Users?**: Enterprises may use certain service accounts that are not part of a SAML 2.0 federation. If that is the case and if the servers using those service accounts need to authenticate, it may not be possible to use this SAML 2.0 vendor and alternative solutions may be needed. The available options are **Yes** or **No**.

- **Is SAML 2.0 Federation Accessible Over the Internet for Remote Users?**: If the remote users are unable to access the SAML 2.0 federation over the internet, then end user authentication using this option may not work. The options are **Yes** or **No**.

- **Is Integrated Windows Authentication (IWA) In Use?**: IWA allows end users who log into their Windows computers to allow those Windows credentials to authenticate into other applications using Kerberos or **New Technology LAN Manager (NTLM)**. The options are **Yes** or **No**.

- **What is your User Directory Type?**: The options are **Active Directory** and **OpenLDAP**.

- **How Many Active Directory Forests Are There?**: The options are **AD Not in Use**, **1**, **2**, and **3 or more**.

Click **Next** to proceed further.

Zscaler Private Access (ZPA) POC

If the enterprise has already done a **proof of concept (POC)** with Zscaler with an ZPA Connector, choose **Yes** for the question **Was ZPA Connector deployed in POC?**. If not, choose **No**. This is useful to know if the enterprise chooses to keep the deployed connector and build on the top of the POC or if it wants to start fresh.

Click **Next** to continue.

ZPA – sites hosting internal apps

To help you decide on the location of the app connectors (whether they're on-premises or in the public cloud), it is necessary to understand the top locations where the enterprise private applications are being hosted:

- **Location Name**: The name of the location. This could be a nickname for an enterprise location, such as "New York Data Center."

- **Location City/State/Country**: List the city, state, and country of this location.

- **Approx Peak Bandwidth Seen**: List the peak bandwidth seen at this location. This is helpful for understanding how many app connectors are needed to support the applications being hosted at this location.

- **Traffic Routing**: If this location has a local internet breakout, choose the **Location Is Directly Connected to Internet** option. If, on the other hand, the local traffic is being backhauled, choose **Location Backhauls Internet Traffic**.

- **Available Infrastructure**: Recall that the app connectors are virtual machines and that as such, they need virtualization infrastructure. Popular options are **VMWare**, **AWS**, and **Azure**.

Click on **Add Another** if your enterprise has more than one main location hosting the private applications. Once done, click **Next** to proceed further.

ZPA interoperability

Recall that some types of applications do not work with Zscaler, so they need to be bypassed from Zscaler and need to take the internet path:

- **Are IT supported VoIP Services/Servers (SIP, Lync, Jabber, Skype) Publicly Accessible from the Internet?**: The options are **Yes (Best Practice)**, **No**, and **Not in Use**.

- **Is Distributed File System (DFS) In Use?**: The options are **Yes** or **No**.

Click **Next** and then **Submit Design Questionnaire** on the next page to complete the question set process and email the results. Notice that the ZPA question set is significantly shorter than the ZIA question set.

Inventory of existing applications

Let's examine the migration process in more detail. This migration process is specific to each enterprise and there is no one-size-fits-all approach. The first step is to inventory the existing enterprise applications and see which ones are a candidate for migration to ZPA. A questionnaire such as the following could be sent to the various teams that manage these applications:

- **Application Name**: The name of the application.
- **Location of the Application**: This could be on-premises, in a private cloud, or in a public cloud.
- **Type of the Application**: Is this a web application or an application that runs off a vendor-provided software program?
- **Volume of Application Requests**: What is the total number of transactions per hour or per day from the end users?

Based on these details, we can identify whether this application is a candidate for browser access and the number of app connectors needed to handle the volume of user requests. It also helps in creating application segments and server groups by grouping similar types of applications.

Discovering end user access

Once the applications have been inventoried, the next step is to understand end user access. How are the end users provisioned access to a certain application? Is it based on the user's job title, department, or location?

Based on the answers to these questions, an access policy can be formulated. Some examples are listed as follows:

- Employees in the Human Resources department can be granted access to Payroll Applications.
- Contractors are explicitly denied access to the Employee Benefits Applications.
- C-Suite Management is allowed access to the internal company document SharePoint.

These are just a few sample examples. This step of the discovery process usually takes a while as existing access policies are questioned if they are still relevant and if they need to be changed or standardized.

Pilot rollout

Once the ZPA administrators have configured the ZPA portal, configured and stood up the app connectors, and then connected them to the application servers using application segments and server segments, the ZPA infrastructure is ready for the pilot users.

Friendly pilot end users are identified by the various teams and passed on to the ZPA migration team. The migration team then provisions these pilot users in the ZPA admin portal and rolls out the ZCC app to their end devices. The pilot users will then attempt to use the same applications that they have been using, but now using the ZPA solution.

Any potential issues that are encountered by those pilot end users are passed on to the ZPA migration team. The migration team then troubleshoots these issues as quickly as possible and adjusts their migration process, so that these issues do not recur when more users are on-boarded to the ZPA solution.

It is also possible to apply the same pilot process to the applications being migrated to ZPA. This means that the enterprise may decide to only migrate a few chosen applications to ZPA during the pilot phase.

Expanded rollout

Once enough pilot users have been using their daily applications through the ZPA solution for a time duration (a few days to a few weeks) deemed sufficient by the senior management, the ZPA migration team typically gets the approval to onboard a larger set of end users.

As the rollout expands across the enterprise, the migration team needs to carefully monitor the peak load on the app connectors and log servers and make sure they can handle the capacity of the new user traffic. If any bottlenecks are observed, the migration team needs to add additional capacity quickly before onboarding more end users. At the same time, the migration team should be ready to resolve any end user issues quickly as the expanded rollout is happening.

Final rollout

Once the expanded rollout has been successful and stable for an acceptable duration of time, it is time to include all the remaining end users and applications for ZPA migration. Once all the end users and applications have been migrated to ZPA, a phased decommission of the pre-ZPA infrastructure can proceed. This completes the enterprise ZPA transformation.

Once you are done architecting and implementing your customized ZPA solution, the next step is to create a unified troubleshooting approach. Let's explore that now.

Troubleshooting your ZPA solution

Once the ZPA solution has been implemented across the enterprise, it is necessary to have a consistent troubleshooting guide to resolve issues seen by the end users during steady state operations.

Let's look at the various types of ZCC app issues that could be seen by the end users and learn how to troubleshoot them.

Unable to access a service due to a captive portal error

When an end user is trying to access the service from a public location such as a coffee shop or an airport, the user usually needs to log into a captive portal after agreeing to the **acceptable use policy (AUP)**. If the end user gets a captive portal error, they should do the following:

1. Ask the user if they have logged into the captive portal.

2. If the user has not completed this step, ask them to accept the AUP and log into the captive portal.

3. Click **Retry** on the ZCC app.

If issues keep occurring, engage higher-tier support where the ZCC app timeout value and the fail-open settings can be checked in the ZCC app portal.

Unable to access a service due to a network error

When an end user runs into this problem, follow these troubleshooting steps:

1. Ask the user if they are connected to the network using a wired or wireless connection.

2. Next, make sure the network adapter for either the wired or wireless connection is not disabled.

3. If the network adapter is disabled, have the user right-click on it and enable it.

4. Once the network adapter shows up as enabled, click **Retry** on the ZCC app.

5. If any issues remain, try to reinstall the device driver for the network adapter.

If the previous step still does not resolve the issue, it could be a hardware issue. Escalate the issue to the appropriate support team.

Unable to access a service due to an internal error

The ZCC app experiences an internal error when it has a problem with the internal sockets. Follow these steps to troubleshoot this issue:

1. If the end user runs into this problem, have the end user click **Retry** on the ZCC app after a few minutes.

2. If any issues remain, ask the end user to reboot the computer. This allows the ZCC app to perform a new socket mapping.

If the issue remains, escalate to the higher tier support.

Unable to access a service due to a connection error

This error usually means that the ZCC app is unable to reach the Zscaler PSE. Follow these steps to troubleshoot:

1. Ask the user to try to perform a `ping` or a `traceroute` to `www.google.com` and the Google DNS IP address of `8.8.8.8`.

2. If the test fails, check why the end user does not have basic network connectivity.

3. If the test is successful, ask the end user to click on **Retry** on the ZCC app.

If the issue remains, escalate this to the higher-tier support to check whether any firewall or routing rules are blocking the user's access to the Zscaler PSE.

Unable to access a service due to a local FW/AV error

This error usually means that the ZCC app is being blocked by a local firewall or an antivirus application. Follow these steps to troubleshoot:

1. Ask the end user if there was a recent update for the antivirus software on their device.

2. Check if any other users are experiencing the same issue.

3. As a test, temporarily turn off the antivirus and/or the local firewall on the computer and have the user click **Retry** on the ZCC app.

If this test works, engage higher-level support to have them whitelist the ZCC app in the antivirus and/or the local firewall software.

Unable to access a service due to a driver error

This issue is usually caused due to a driver installation failure, which causes the ZCC app to be unable to turn on the tunnel interface. Follow these steps to troubleshoot:

1. Ask the end user to click on the **More** icon on the ZCC app and click on **Repair App** under the **Troubleshoot** section.

2. If the user does not have that option enabled, escalate this to the higher-level support so that they can turn on this option for the users.

3. Once the repair process completes, have the user click on Retry on the ZCC app.

If the issue remains unresolved, escalate it to the next level.

Unable to access a private application/service due to an unauthenticated error

By default, the user's authenticated session expires after 7 days and the user must reauthenticate periodically to maintain access to the private applications. Follow these steps to troubleshoot:

1. Ask the end user to authenticate again and access the application again.

2. If the user authentication process fails, check why (it could be a wrong password, their password may have expired, and so on).

If trouble remains, escalate the issue to the next level.

Unable to access a private application/service

In this case, the user simply cannot access the private application. Follow these steps to troubleshoot:

1. Ask the end user to attempt to access the application and note down the exact error message.

2. Ask about the last time the application was working for the user. If it never worked for the user, perhaps the access was never provisioned.

3. Check if the user can access any other private applications. This helps determine if it is a per-application access issue or a user access-level issue.

4. If the end user never requested access to the application, have them go through the access request submission process and have them retry when access is approved and provisioned.

If the issue persists, engage higher-level support.

Unable to access any application/service

If the end user cannot access any private application at all, there could be two issues:

-   The user might not have basic network connectivity. In this case, perform basic network troubleshooting and resolve network connectivity issues and have the user click **Retry** on the ZCC app.

-   The second reason might be the user was never provisioned for any private applications. Check this with the ZPA administration team.

If you are unable to resolve this issue, escalate to next-level support.

The end user needs to authenticate into the ZCC app using the IdP in use by the enterprise. Let's look at some of the types of SAML authentication errors that may occur.

Unable to authenticate due to a SAML transit error

SAML transit errors are temporary in nature and usually clear on their own. Follow these troubleshooting steps:

1. Check if the end users are seeing one of these error codes: E5503, E5507, E5508, E5611, E5612, E5614, E5619, E5623, E5629, A002, A003, A00C, A00D, A00E, A011, A019, A023, A029, or A02A.

2. Ask the end user when was the last time authentication was working.

3. Check how many other end users are experiencing the same issue.

4. Ask the user to close their browser session and retry authentication after a while.

If issues remain, escalate to next-level support.

Unable to authenticate due to a SAML account error

SAML account errors are experienced by end users when there is an issue with their user accounts. Follow these troubleshooting steps:

1. Ask the end user to attempt to authenticate into the ZCC app and check if they are seeing one of these error codes: `E5616`, `E5621`, `E5624`, `E5628`, or `A010`.

2. Note down the username of the end user and check if that account exists on the IdP or on the ZPA admin portal.

3. If the username does not exist in either or both, please engage the appropriate support team.

4. If the username is found, check if it is in a disabled state. If so, get the user's account reenabled through proper procedure.

5. Finally, check if the auto-provisioning feature is enabled on the ZPA Admin Portal.

If the issue persists, engage higher-level support.

Unable to authenticate due to a SAML format error

User authentication into the ZCC app needs to be done with an email address. Follow these troubleshooting steps:

1. Ask the end user to attempt to authenticate and capture the error code. See if the error code is `A021`.

2. Check the format of the username being used by the user for authentication.

3. Verify that the user ID and the domain are correct and that they match the configuration in the ZPA Admin Portal.

If the issue remains, engage higher-level support.

There are many types of error codes that are displayed by the ZCC app to troubleshoot these scenarios. These are provided in links in the *Further reading* of this chapter.

Summary

In this chapter, we learned how to walk through the ZPA question set and make the right decisions when it comes creating a customized ZPA solution for our own enterprise. We also learned how to troubleshoot ZPA end user issues during everyday operations.

Questions

As we conclude, here is a list of questions for you to test your knowledge regarding this chapter's material. You will find the answers in the *Assessments* section of the Appendix:

1. App connectors should always be deployed in pairs for redundancy.

    a. True

    b. False

2. ZPA app connectors cannot be deployed in public clouds.

    a. True

    b. False

3. All enterprises should use a one-size-fits-all ZPA solution.

    a. True

    b. False

4. Most ZPA issues seen by end users are Zscaler issues.

    a. True

    b. False

Further reading

- ZPA Authentication Errors: `https://help.zscaler.com/z-app/zscaler-app-zpa-authentication-errors`
- ZPA Session Status Codes: `https://help.zscaler.com/zpa/about-zpa-session-status-codes`

Assessments

In the following pages, we will review all practice questions from each of the chapters in this book and provide the correct answers.

Chapter 1 – Security for the Modern Enterprise with Zscaler

1. a and b
2. All the above
3. All the above
4. False
5. False
6. True

Chapter 2 – Understanding the Modular Zscaler Architecture

1. Central Authority, Public Service Edge, Nanolog Cluster, Sandbox
2. Central Authority
3. False
4. False
5. 6 months
6. True

Chapter 3 – Delving into ZIA Policy Features

1. All the above

2. False

3. All the above

4. True

Chapter 4 – Understanding Traffic Forwarding and User Authentication Options

1. True

2. False

3. Hosted DB, LDAP, SAML, Kerberos

4. False

Chapter 5 – Architecting and Implementing Your ZIA Solution

1. True

2. Configuration, Production Rollout, Pilot Rollout, Planning

3. False

4. True

Chapter 6 – Troubleshooting and Optimizing Your ZIA Solution

1. False

2. True

3. All the above

Chapter 7 – Introducing ZTNA with Zscaler Private Access (ZPA)

1. Yes
2. No
3. False
4. False

Chapter 8 – Exploring the ZPA Admin Portal and Basic Configuration

1. True
2. All the above
3. None of the above
4. False

Chapter 9 – Using ZPA to Provide Secure Application Access

1. All the above
2. None of the above
3. False

Chapter 10 – Architecting and Troubleshooting Your ZPA Solution

1. True
2. False
3. False
4. False

Other Books You May Enjoy

If you enjoyed this book, you may be interested in these other books by Packt:

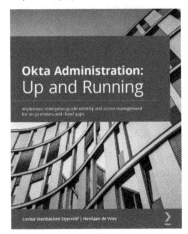

Okta Administration: Up and Running

Lovisa Stenbäcken Stjernlöf, HenkJan de Vries

ISBN: 978-1-80056-664-4

- Understand different types of users in Okta and how to place them in groups
- Set up SSO and MFA rules to secure your IT environment
- Get to grips with the basics of end-user functionality and customization
- Find out how provisioning and synchronization with applications work
- Explore API management, Access Gateway, and Advanced Server Access
- Become well-versed in the terminology used by IAM professionals

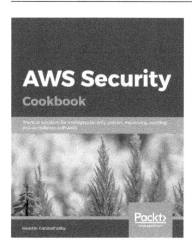

AWS Security Cookbook

Heartin Kanikathottu

ISBN: 978-1-83882-625-3

- Create and manage users, groups, roles, and policies across accounts
- Use AWS Managed Services for logging, monitoring, and auditing
- Check compliance with AWS Managed Services that use machine learning
- Provide security and availability for EC2 instances and applications
- Secure data using symmetric and asymmetric encryption
- Manage user pools and identity pools with federated login

Packt is searching for authors like you

If you're interested in becoming an author for Packt, please visit authors. packtpub.com and apply today. We have worked with thousands of developers and tech professionals, just like you, to help them share their insight with the global tech community. You can make a general application, apply for a specific hot topic that we are recruiting an author for, or submit your own idea.

Leave a review - let other readers know what you think

Please share your thoughts on this book with others by leaving a review on the site that you bought it from. If you purchased the book from Amazon, please leave us an honest review on this book's Amazon page. This is vital so that other potential readers can see and use your unbiased opinion to make purchasing decisions, we can understand what our customers think about our products, and our authors can see your feedback on the title that they have worked with Packt to create. It will only take a few minutes of your time, but is valuable to other potential customers, our authors, and Packt. Thank you!

Index